海味乾貨家常菜

用最簡捷的方法處理，烹調每天的美味佳餚

Dried Seafood Dried Goods Homestyle Recipes

Using the Easiest and Quickest Way of Preparing and Cooking at Home

當海味、乾貨的構思浮現在腦袋，就着手去探討它們的處理方法；我是説認真的去探討、研究……

因為平時處理海參、花膠，都用自己想當然的方法，膽粗粗就做了，雖然不算差，但要白底黑字呈獻給讀者，就得格外小心了。

海味、乾貨可以做成大菜，奉為桌上皇牌，也可以毫不顯眼的成為「茄喱啡」，明知它們不易處理，卻一心要用簡易的處理方法帶進您的府上，讓您也掌握到烹調它們的竅門。是皇牌也好，是「茄喱啡」也好，都可以得心應手的為您贏得掌聲。

記住，我們做的，是融合世界御廚、鮑魚大王（楊貫一）及徐師奶（黃淑儀）的方法，絕對令您滿意！

I went into some serious researching to gather detailed information about dry goods and seafood in order to do the title of the cookbook justice!

Dry goods and seafood can often be presented as the main star attractions at banquets, yet they can be the inconspicuous extras as well.

Realizing that they are not that simple to handle, I want to introduce to you some easy ways to prepare them at home. It is for you to grasp the secrets of treating them, be they the main dishes or the accompaniments.

Remember, what we have here is the combined methods of the world renown King of Abalone, Yeung Koon Yat and Mrs home chef Tsui, aka Gigi Wong – guaranteed for satisfaction!!

目錄 Contents

內附 QR 碼，手機一嘟即時去片！

海味 Dried Seafood

乾貨 Dried Goods

海味

為鮑魚、海參、花膠、蠔豉、瑤柱、大地魚、蝦乾……配搭合適的配料、調味料和烹調法，將海味鮮香、濃郁的特性發揚光大，誘發你的食欲。

Dried Seafood

Featuring the right ingredients, treatments and cooking methods to enhance the aroma and freshness of dried abalone, dried sea cucumbers, dried scallops, dried oysters, dried shrimps and a host of dried seafood - guaranteed to whet your appetite!

海味・鮑魚 ABALONE

在香港，在中國，在東南亞，甚至在全世界的飲食之都，只要提起鮑魚，都知道香港的楊貫一。

果然是阿一鮑魚，天下第一！名不虛傳也！

但，更難得的是這位鮑魚之父絕不吝嗇，他將自己炮製鮑魚的秘方公開，甚至好細心地提示每個要點。

當然，能否做到一哥的水準，就要看你的造化了；例如：

- 你的材料達到水準嗎？
- 你的用具齊全嗎？
- 你的火候控制得宜嗎？
- 最重要是你的愛心、耐性合格嗎？

In all the food capitals of the world including Hong Kong, China and South East Asia, the name of Ah Yat is synonymous with abalone.

Without doubt, Ah Yat's abalone is deservedly the world number One!

The Father of Abalone has generously and openly shared his secrets for preparing abalone in detail with us.

Obviously, whether we can achieve his level of expertise is very much dependent on our grasp of his tips. For example,

- How do you select the ingredients correctly?
- Is your set of utensils complete?
- Is the cooking temperature controlled well?
- Or, do you have the patience required for it?

Ah Yat Shows the Way

一哥教路整鮑魚

Utensils

用具

- 瓦煲 ....1 個
- 竹筷子 ....2 對
- 竹簍 ....2 個
- 布袋 ....2 個

- 1 claypot
- 2 pairs bamboo chopsticks
- 2 bamboo sheets
- 2 cloth bags

材料

- 乾鮑 ....600 克
- 肉排 ....900 克
  切開 4 件
- 老雞 ....1 隻
  切開 4 件
- 水 .... 到面即可

- 600 g dried abalone
- 900 g pork ribs :
  cut into four rows
- 1 mature chicken :
  cut into four
- water enough to cover all the ingredients

Gravy

獻汁

- 鮑魚汁 ....1 湯匙
- 火腿汁 ....1 湯匙
- 鵝掌汁 ....2 湯匙

- 1 tbsp abalone sauce
- 1 tbsp Jinhua ham sauce
- 2 tbsps goose web sauce

MeThod
做法

❶先將鮑魚用水浸八小時，然後洗擦乾淨。
❷把肉排及老雞出水，分別放在布袋內，待用。
❸先把筷子放在瓦煲底，然後放竹篷，再把肉排排好，上面放鮑魚，再放上老雞，注入過面水。
❹中火煲約 10 小時，每小時觀察一次，若水不夠，則加雞湯一次。測試鮑魚是否已煲腍，用牙籤容易插入兼有鮑魚香即可。
❺最後過程是收汁，把鮑魚全部取出，把剩下的汁液用小火煲至稠。
❻燜好的鮑魚在攤凍後，用保鮮紙個別包好，再放進密封盒，外用保鮮紙包裹，置冰格可保存 3 個月。
❼燜鮑魚剩下的湯汁，撇油後，另盒裝起。
❽享用前，將鮑魚汁和已解凍的鮑魚用慢火加熱，下獻汁，用少許老抽調色，打生粉獻即可。

❶ Firstly, soak dried abalone in water for 8 hours, then wash and wipe clean.
❷ Scald both the pork ribs and chicken. Store them separately in the cloth bags for use.
❸ Put the bamboo chopsticks at the bottom of the claypot; lay bamboo sheets over them. Place the pork ribs on it, then arrange the abalone over it, finally leave the chicken on top of them.
❹ Braise for 10 hours over medium heat. Do check it every hour to see whether there is sufficient water in the pot. If not, add some chicken soup. To check whether the abalone is done, prick it with toothpick. If it is done with ease and the abalone is fragrant, it is cooked.
❺ The last step is to reduce the sauce. Remove all the abalone from the pot, leave only the sauce to boil over low heat till it thickens.
❻ After cooling off the braised abalone, wrap them individually in cling film before storing them in sealed boxes. Seal again with cling film to place in freezer for up to 3 months.
❼ Scrape off the oil of the residual sauce from braising the abalone and keep it in a separate box.
❽ Before serving, heat up the abalone sauce and the defrosted abalone over low heat. Pour the gravy on it and enhance the color by seasoning it with dark soy sauce, and thicken it with cornflour solution.

＊＊ 鮑魚的品種很多，常食的鮑魚分類就有窩麻、吉品、細乾鮑及罐頭鮑。
日本大間出產的窩麻，較受老人家歡迎，因為較軟滑。而產地在岩手縣的吉品，因有咬口，所以甚受年青人的歡迎。
＊＊罐頭鮑主要是易於處理，只要連罐放在煲內，浸在水裏煲 8 小時，凍後開罐，切片，埋獻即可。
＊＊ 記住燜鮑魚時不可有鹹味，即是說：燒豬骨或火腿等帶鹹味的配料就不宜加入同燜。
＊＊ 鮑魚通常要燜 10-14 小時，視乎大小而定，細隻乾鮑燜的時間可以縮短。
20 頭以下吉品要煲 16 小時；20 頭以上要煲約 20 小時；網鮑體積較大，要煲 3 日，每日 8 小時。

** Among the many varieties of abalone, the common ones are Wo Ma (Oma), Kippin, small dried abalone and canned abalone. Oma, produced in Oma, Japan, is popular amongst the older group as they are softer and smoother. However Kippin which is produced in Yoshihama, Japan, is favored by the young for its chewy texture.
** Canned abalone is easy to handle. All you need to do is place the entire can in a pot, soak it in water to boil for 8 hours. While it cools, open the can, slice it and pour sauce on it to serve.
** Remember though, do not mix in salty flavor while braising abalone, in other words, do not add salted foods such as roasted pork bones or ham together in the pot.
** Depending on its size, abalone generally needs 10-14 hours to braise. The cooking time can be reduced if abalone is smaller. Those below 20 heads Kippin would require 16 hours to braise; above 20 heads, 20 hours. Japanese Amidori abalone is larger in size; hence it would need 8 hours per day for 3 days to braise.

TIPS 享用貼士

享用時，宜先用刀义將鮑魚打直切開，再橫切成片蘸獻汁享用。

Cut open abalone with a knife in a vertical line, and then slice it into thin pieces to dip in the gravy for serving.

海味・海參

DRIED SEA CUCUMBERS

挑選・SELECTING

一般海參以乾身、結實、肉厚、完整無缺為佳。

The good sea cucumbers are the dried, firm, fleshy ones which are also intact.

處理・TREATING

海參要先浸 2 天，早晚換水；放在過面水中加熱煲滾 30 分鐘，熄火後不揭蓋，讓海參在煲內焗至涼。

翌日會發現海參發大、體質變軟，可用剪刀剪開海參，取出體內腸臟、沙粒等髒物，再沖洗乾淨，是為第一步。

完成的第一步，只可以說是初步處理，我們仔細分辨它的軟硬程度，分別裝在三文治袋內封好，更標簽清楚它的軟、硬度，放進冰格，用時取出解凍即可進入第二步——加工。
加工，是表示此時此刻的海參尚未完成處理，未可以入饌，視乎軟硬程度再氽水 3-5 次，但，為甚麼我建議在第一步後就入雪櫃儲存呢？那是因為海參經浸泡、氽水後，體內的小孔擴張，一旦放入冰格，小孔的水分都變成冰，解凍後海參就更軟了，所以，如果在完全處理後放入冰格，到應用時， 海參就糜爛不堪，浪費了。

要做第二步之前，先將海參解凍，軟的只需要氽水一次，煨一次即可，硬身的則需要再氽水 3-5 次，才可煨過採用。完成第二步的海參要立刻烹調，不宜再放入冰格。

** 氽水是令海參變軟，而氽水用的煲不可有油漬，否則海參容易溶化。

處理 • TREATING

Soak dried sea cucumbers in water for 2 days. Change the water in the morning and evening. Put them fully covered in a pot of boiling water to boil for 30 minutes. After the flame is switched off, leave the cover intact and let the sea cucumbers remain till cool.

The next day, the dried sea cucumbers would be enlarged and softened. Use a pair of scissors to cut them open, discard the innards and wipe any impurities off. This is the first step.

Observe carefully their degree of softness. Separate them in sandwich bags to seal. Mark on the bags their degree of softness. Leave in the freezer. When using it, defrost it, and on to the second step – processing them.
Processing means the dried sea cucumbers are not treated for cooking yet. Depending on their softness, scald them 3-5 times again. However, why did I suggest freezing them after the first step? It is because after soaking and scalding, small holes in dried sea cucumbers expand. When frozen, the water in the holes would turn into ice. They will be soft when defrost. Therefore, if the dried sea cucumbers are frozen after being processed, they will be too mushy for use.

Before attempting the second step, scald and simmer the softer ones only once; scald the harder ones 3-5 times more before simmering them for use. After the second step, use them immediately.

** Scalding makes the dried sea cucumbers softer. Please make sure the pot used is oil free or the sea cucumbers will dissolve.

煨法 • SIMMERING

烹調前須煨過海參，就可以辟去腥味。

用2湯匙油爆香薑2片、葱2條，濽酒2湯匙，再加入海參，煨煮5分鐘後，將海參取出即可入饌。

Before using them for cooking, simmer them first to get rid of the odor.

Fry 2 slices of ginger and 2 stalks of spring onions in 2 tbsps of oil till fragrant. Pour 2 tbsps of wine on the side of wok. Put in dried sea cucumbers. Simmer for 5 minutes; scoop up the dried sea cucumbers. They are now ready for use.

用途 • USAGES

最常見的煮法是燜、燒、炒、燉。

Commonly used for braising, stir-frying and stewing.

Stir Fried Sea Cucumbers with Pepper

辣椒炒海參

InGreDienTs 材料

- 海參 ....400 克
  處理好，切塊
- 梅頭豬肉 ....50 克
  切片，用 1 茶匙生抽、
  1 茶匙粟粉及少許胡椒粉略醃
- 尖嘴青椒 ....100 克
  先洗淨，切塊
- 葱白 ....2 條
  切度
- 蒜頭 ....2 粒
  拍扁
- 薑 ....2 片

- 400 g sea cucumbers : treated and sectioned
- 50 g pork loin: sliced and marinated with 1 tsp soy sauce ,1 tsp cornflour and a pinch of pepper
- 100 g green pepper : washed and sectioned
- 2 sprigs white part of spring onion : sectioned
- 2 cloves garlic : flattened
- 2 slices ginger

SeaSonIng 調味料

- 麵豉醬 ....3 湯匙
- 紹酒 ....1 湯匙
- 糖 ....2 茶匙

- 3 tbsps fermented bean sauce
- 1 tbsp Shaoxing wine
- 2 tsps sugar

MeThod 做法

❶用 1 湯匙油爆香蒜頭、葱白，加入梅頭豬肉，灒酒，下青椒，炒至肉熟，盛起。

❷再用 2 湯匙油爆香薑片及麵豉醬，下海參，燜至溢出香味，再把 [1] 加入同炒，最後加糖炒勻即可。

❶ Fry the garlic and spring onion in 1 tbsp of oil till fragrant. Add the pork slices. Pour wine on the sides of the wok, put the green pepper in to fry till meat is cooked. Scoop up.

❷ Fry the ginger slices and fermented bean sauce in 2 tbsps of oil till fragrant, add the sea cucumbers in to braise till fragrant. Put in (1) and stir together; lastly add sugar to stir well. Serve.

Dried Sea Cucumbers and Chicken Rice

海參鮮雞飯

InGreDienTs 材料

- 海參 ....150 克
  處理好，切塊
- 雞 .... 半隻
  切塊
- 薑米 ....1 湯匙
- 米 ....3 杯
  洗淨，瀝去水分
- 水 ....2 杯

- 150 g sea cucumbers : treated and cut into pieces
- 1/2 chicken : cut into pieces
- 1 tbsp diced ginger
- 3 cups rice : washed and drained
- 2 cups water

MariNade 醃料

- 鹽 .... 半茶匙
- 粟粉 ....1 茶匙
- 生抽 ....1 湯匙
- 薑汁酒 ....1 湯匙

- 1/2 tsp salt
- 1 tsp cornflour
- 1 tbsp light soy sauce
- 1 tbsp ginger wine

MeThod 做法

❶將雞醃 1 小時，用 2 湯匙油爆香薑米及雞至半熟，待用。
❷將米與水放入電飯煲內，按掣。
❸電掣跳起後，放入海參及雞，拌勻，再按電掣。
❹電掣再跳起後，不要開蓋，讓飯焗 20 分鐘即成。
❺可酌量加生抽調味。
＊＊因雞在烹調過程時會溢出雞汁，所以煲飯用 2 杯水已足夠。

❶ Marinate the chicken for 1 hour, fry the chicken and diced ginger in 2 tbsps of oil till the chicken is half cooked. Set aside.
❷ Place the rice and water in an electric rice cooker. Switch it on.
❸ When it is switched off, put the dried sea cucumbers and chicken in, mix well, and press the "on" switch again.
❹ When the switch is turned off, do not open the lid, allow the rice to cook for another 20 minutes.
❺ Season it with some soy sauce.
**Using 2 cups of water for cooking rice is sufficient as chicken oozes juice during cooking.

海味・花膠

FISH MAW

挑選 • SELECTING

一般家庭食用花膠，不需要買太名貴的，就算是普通的花膠筒，只要烹調得宜，也很不錯。
宜挑選厚身（當然一分價錢一分貨）、顏色金黃、乾身的；如在燈光下，發現花膠有特別深色的部分，不是剔透的，那表示在曬製過程中積聚水分，不宜久存，就不要買了。

Buy the standard fish maw for home use. They need not be costly because, if cooked well they can be just as tasty.
Pick those thicker bodied, golden yellow colored and dried ones. Under the light, if the fish maw shows sections which are darker in color, do not select it because it shows there is water retention during the drying process. The affected fish maw cannot be stored long.

處理 • TREATING

先用清水浸花膠一天，沖洗乾淨。煮一煲滾水，放入花膠滾 10 分鐘，熄火，焗 6-10 小時或過夜，取出花膠，洗淨。如想即刻烹調，可將部分花膠放入薑葱滾水內氽水，去腥味，盛起即可烹調。至於剩下的花膠，可放入三文治袋內，封口，貯存在冰格內可保鮮半年。

Soak fish maw in fresh water for one day. After rinsing them, boil a pot of water and boil it for 10 minutes. Switch off the flame but leave it in the pot for 6-10 hours or overnight. Remove and wash them. For immediate use, place some fish maw into boiling water with ginger and spring onion to scald. This is to get rid of the odor. Scoop and use them for cooking. Store the remaining fish maw in sandwich bags, seal and put in freezer to keep fresh for up to 6 months.

用途 • USAGES

宜燜、燉、煮、炒、煲湯、煲糖水。

For braising, double-steaming, simmering, frying, boiling soup and dessert.

挑選 • SELECTING

宜選有淡淡鮮香、爆得鬆浮的魚肚，如爆不透，煮時會瀉身。

Choose the ones which are puffed up well with some light aroma. If they are not inflated well, they are shapeless.

處理 • TREATING

將魚肚浸軟；煲內放入水、1 片薑、1 棵葱和 2 滴白醋，放入魚肚汆水，撈起，擠乾水分即可烹調。
這種魚肚很易浸軟，不需要預早幾日處理。下白醋的作用是可辟腥。

Soak fish maw till they are soft. Then put 1 piece of ginger, 1 sprig of spring onion, 2 drops of white vinegar and water in a pot to scald them. Scoop them up, squeeze them dry and they are ready for cooking.
This type of fish maw softens easily; hence it is unnecessary to treat them well in advance. The purpose of adding white vinegar is to rid it of its odor.

用途 • USAGES

宜作湯羹、燜煮，它那鬆浮的質感吸味一流。

Good for making thick soups and braising. Its soft and puffed up texture absorbs flavor well.

海味 • 砂爆魚肚
SAND-HEATED FISH MAW

Double-steamed Fish Maw

燉花膠

InGreDienTs 材料

- 花膠 ....1 條
  處理好
- 甘草 ....1 片
- 姬松茸 ....3 個
  浸軟，剪去硬蒂
- 陳皮 ....1 角
  浸軟，刮去瓤
- 滾水 ....2 碗

- 1 fish maw : treated
- 1 slice licorice
- 3 agaricus : softened in water and stems discarded
- 1 piece dried tangerine peel : softened in water, pith removed
- 2 bowls boiling water

MeThod 做法

❶將所有材料放入燉盅內，用中火燉 3 小時即可。

＊＊這燉湯有減低血糖、治糖尿病的療效

❶ Place all the ingredients in a double-steaming pot and steaming over medium heat for 3 hours before serving.

** The soup is good for lowering blood sugar and curing diabetes.

Fish Maw Sweet Soup

花膠甜湯

InGreDienTs 材料

- 花膠 ....1 塊
  處理後，切絲
- 蓮子 .... 半杯
  用水浸 2 小時後，再用高火「叮」10 分鐘
- 桂圓肉 ....1/3 杯
  沖洗乾淨
- 無花果 ....10 粒
  沖洗乾淨
- 陳皮 ....1 角
  用水浸軟，刮去瓤
- 薑 ....1 片
- 冰糖 ....2 湯匙
- 水 ....6 杯

- 1 piece fish maw :
  treated and shredded
- 1/2 cup lotus seeds :
  soaked in water for 2 hours, then microwaved on high for 10 minutes
- 1/3 cup shelled dried longans :
  washed
- 10 dried figs : washed
- 1 piece dried tangerine peel :
  softened in water, pith removed
- 1 slice ginger
- 2 tbsps rock sugar
- 6 cups water

MeThod 做法

❶煲滾水，放入所有材料，煲約 1 小時，就可享用。

❶ Place all the ingredients into the boiling water to cook for 1 hour. Ready to serve.

Stir Fried Fish Maw

醬燒花膠

InGreDienTs
材料

- 花膠 ....1 條約 500 克
  處理好，切方形
- 西蘭花 ....300 克
  切小朵，洗淨

- 1 about 500 g fish maw :
  treated and cut into pieces
- 300 g broccoli :
  cut into florets and washed

SeaSonIng
調味料

- 海鮮醬 ....3 湯匙
- 薑米 ....2 湯匙
- 蒜茸 ....2 湯匙
- 紹酒 ....1 湯匙

- 3 tbsps Hoisin sauce
- 2 tbsps diced ginger
- 2 tbsps minced garlic
- 1 tbsp Shaoxing wine

MeThod
做法

❶煲滾 2 杯水，加入 1 湯匙油、1 茶匙鹽，下西蘭花，焯 2 分鐘，撈起沖凍水，保持翠綠。
❷西蘭花排在碟邊，待用。
❸用 2 湯匙油爆香薑米、蒜茸及海鮮醬，倒入花膠爆炒，濳酒，燒至海鮮醬均勻地附在花膠上，即可盛起，放在中央即成。

❶ Bring 2 cups of water to the boil. Add 1 tbsp of oil and 1 tsp of salt to blanch the broccoli for 2 minutes. Scoop up and splash them with cold water to preserve the green color.
❷ Arrange them on the side of a plate. Set aside.
❸ Fry the diced ginger, minced garlic and the Hoisin sauce in 2 tbsps of oil till fragrant. Add fish maw to stir quickly, pour wine on the sides of the wok. Cook till the sauce stays evenly on the fish maw. Scoop and place them at the center of the broccoli.

Angled Loofah and Sand-heated Fish Maw in Fish Soup

勝瓜蝦乾浸魚肚

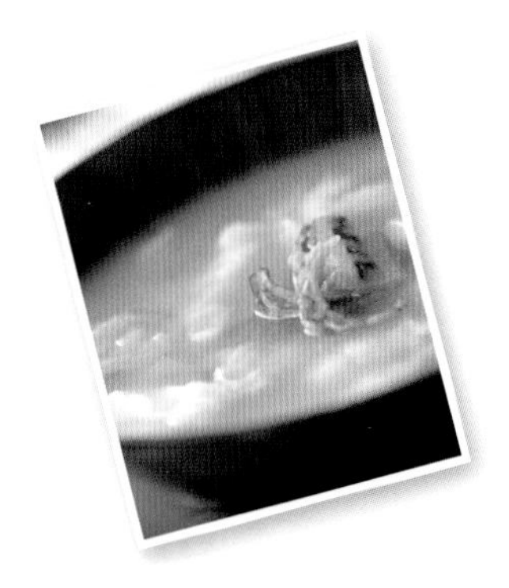

InGreDienTs 材料

- 勝瓜 ....1 條
  刨去粗皮，切件
- 砂爆魚肚 ....38 克
  處理好，切絲
- 蝦乾 ....38 克
  洗淨，浸軟
- 薑 ....1 片
- 水 ....6 杯

- 1 angled loofah : rough skins removed and sliced
- 38 g sand-heated fish maw : treated and shredded
- 38 g dried shrimps : treated
- 1 slice ginger
- 6 cups water

MeThod 做法

❶煲滾水後，加入薑片、蝦乾及魚肚絲滾 5 分鐘，再放入勝瓜滾 1 分鐘，下半茶匙鹽即可享用。

❶ Bring the water to the boil then add the sliced ginger, dried shrimps and fish maw. When the water is brought to the boil again, put loofah in. Boil for 1 minute and season with 1/2 tsp of salt. Serve.

海味・瑤柱 DRIED SCALLOPS

挑選 • SELECTING

宜挑選色澤金黃，手感乾燥，有鮮香味的瑤柱。
Select those which are golden in color; feel dry to the touch and they are fragrant and fresh.

處理 • TREATING

以前我都是用水浸軟瑤柱，但經浸泡的瑤柱色和味都變淡，自從經一哥教路將瑤柱泡油才處理，鮮味得以保存。
將半斤瑤柱用扭乾的濕布抹乾淨，準備小鍋，加約兩吋油，用中火將瑤柱炸一分鐘至微黃即撈起，放在廚房紙上索去油分。
涼後放入三文治袋內，放在冰箱可保鮮半年，用時取出加過面水，放入微波爐用高速「叮」十分鐘，共「叮」三次，每次都要查看水分是否充足。

Previously I soaked scallops in water but they turned light in color and not as tasty. With the tips given by Ah Yat, I now scald them in hot oil first before cooking to preserve the fresh taste.
Clean 300 g of dried scallops with a piece of dry cloth. Add about 2 inches of oil in a small pot; fry the dried scallops over medium heat for 1 minute till they turn slightly golden. Scoop up and place on kitchen paper towel to absorb the oil.
When they are cool, place them in sandwich bags to keep in freezer for up to 6 months. To use them, cover them with water up to the surface. Cook in microwave oven in high speed for 10 minutes. Repeat the cooking three times. For each cooking cycle do check whether the water used is sufficient.

用途 • USAGES

可當零食、菜肴、糕點，如瑤柱扒豆苗、節瓜瑤柱脯、瑤柱白粥、蘿蔔糕等等。
Good to use it in snacks, dishes and cakes such as stir fried dried scallops with bean shoots, steamed hairy gourd with dried scallops and radish cake.

Dried Scallops with Fresh Vegetables

瑤柱鮮蔬

InGreDienTs
材料

- 番茄 ....1 個
  去皮，切塊
- 大芥菜 ....1 個
  洗淨，切塊，用油鹽汆水
- 粟米仔 ....6 條
- 瑤柱絲 ....2 湯匙
  處理好，炸香
- 薑 ....1 厚片
- 雞湯 .... 半杯

- 1 tomato :
  skinned and cut into wedges
- 1 mustard green : washed, sectioned and blanched with oil and salt
- 6 baby corns
- 2 tbsps shredded dried scallops: treated and fried
- 1 thick piece ginger
- 1/2 cup chicken stock

MeThod
做法

❶用 2 湯匙油爆香薑片，放入番茄、大芥菜及粟米仔爆炒，下鹽 1 茶匙，倒入雞湯，加蓋焗 2 分鐘盛起。
❷瑤柱絲散放在鮮蔬上即完成。

❶ Fry the ginger in 2 tbsps of oil till fragrant, add the tomato, mustard green and baby corn in to stir fry. Add 1 tsp of salt, pour the chicken stock in, cover and cook for 2 minutes. Scoop up.
❷ Sprinkle the shredded dried scallops over it to serve.

XO Sauce

XO 醬

InGreDienTs 材料

- 薑 ....300 克
  去皮，磨茸
- 蒜頭 ....300 克
  去皮，磨茸
- 辣椒粉 ....600 克
- 生油 ....600 克

- 300 g ginger :
  skinned and minced
- 300 g garlic :
  skinned and minced
- 600 g chili powder
- 600 g oil

AccomPaniments 配料

- 金華火腿 ....150 克
  汆水，切細粒
- 瑤柱 ....150 克
  處理好
- 蝦子 ....2 湯匙

- 150 g Jinhau ham :
  scalded and diced
- 150 g dried scallops :
  treated
- 2 tbsps dried shrimp roe

MeThod

做法

❶燒熱半鑊油，先放入薑爆炒，再下蒜茸，待兩者都爆香後，小心倒入辣椒粉，即可熄火，待冷。

❷以上是做 XO 醬的基本醬底，加配料後不能久藏，故若要加配料，只需用油爆香配料，再混合部分基本醬即可，情願吃完再混合。

＊＊先爆香薑，後爆蒜，是因為蒜較搶火，易焦；而辣椒粉加入油鑊中，會立刻泛起油泡，所以要特別小心。

＊＊基本醬一定要多油，蓋過面，蒜茸及辣椒粉才可保存長久。

❶ Heat half a wok of oil; stir fry the minced ginger, then add the minced garlic. Carefully add the chili powder. Switch off flame and let cool.

❷ What is done in (1) makes the basic XO sauce. If the accompaniments are added, the sauce cannot be kept for too long. Fry the accompaniments in oil till fragrant to add to the basic sauce. It is best to make in batches when it is required.

** Fry the ginger first before garlic because garlic burns easily. Be careful when chili powder is added to the oil in work as oil bubbles will surface.

** Add enough oil to cover the surface of basic sauce so as to preserve garlic and chili powder longer.

XO Sauce

XO 醬

海味・蠔豉 DRIED OYSTERS

挑選 • SELECTING

以蠔身飽滿、色澤金黃油潤，散發鮮香氣味的為佳品。如蠔身瘦瘠、顏色黯啞、有霉味的為次貨。

Full-bodied, golden yellow, moist and fragrant dried oysters should be picked. The inferior products are shriveled up, dried, darker in color and smells moldy.

處理 • TREATING

生曬蠔豉：用溫水浸軟蠔串，用較剪分開蠔肉和竹籤，再將蠔肉褪離竹籤，如蠔豉面有白灰，可用牙刷擦去，洗淨待用。
一般蠔豉：用溫水浸軟，洗淨即可。

Sundried oysters: soak sundried oysters in warm water. Cut and separate the oyster flesh and bamboo stick with a pair of scissors. Slide the flesh off from the stick. If the surface has white grayish stuff, brush it off with a toothbrush and wash before use.
Standard dried oysters: soak dried oysters in warm water until tender. Wash.

用途 • USAGES

生曬蠔豉多用來蒸、煎、燜，味道甘香；一般蠔豉多用來煲湯、燜等。

Sundried oysters are mainly used for steaming, frying and braising dishes. They are aromatic. The standard dried oysters are for boiling soup and braising.

Dried Oysters in Mirin Sauce

味酥金蠔

InGreDienTs
材料

- 生曬蠔豉 ....12 隻
  處理好，洗淨
- 自製味醂汁 ....1 杯

- 12 sundried oysters :
  treated and washed
- 1 cup homemade mirin sauce

Homemade Mirin Sauce
自製味醂汁

- 味醂 ....2 杯 (500 毫升)
- 生抽 ....2 杯 (500 毫升)
- 冰糖 ....200 克

- 2 cups mirin
  (about 500 ml)
- 2 cups light soy sauce
  (about 500 ml)
- 200 g rock sugar

MeThod
做法

❶用慢火煮滾味醂汁材料，待冰糖煮溶後，即可關火，待涼。

❷將生曬蠔豉放入半杯自製味醂汁內浸至入味 (汁要蓋過蠔豉，醃約 2-3 小時)，取出生曬蠔豉，瀝乾汁液，撲上粟粉，再用慢火煎至金黃，倒入餘下的味醂汁，焖至汁液濃稠即可。

＊＊因為味醂汁的糖分重，一定要用慢火煮及要注意火候。

＊＊每人的口味不同，製味醂汁時一定要試味，再加減糖的份量。

＊＊需要用過面味醂汁浸生曬蠔豉，剩下的味醂汁須放在雪櫃，可保鮮一年。

＊＊除生曬蠔豉外，味醂汁可用來醃雞翼、牛仔骨、銀鱈魚等等。

❶ Boil mirin sauce ingredients over low flame. When the rock sugar dissolves, switch the flame off and cool.

❷ Place the sundried oysters in the 1/2 cup of mirin sauce to soak till they are flavorful (the sauce is to cover all the oysters to marinate for about 2-3 hours). Remove the oysters and drip dry. Coat them with cornflour. Fry them over low heat till golden in color. Pour the remaining sauce in to cook till the sauce thickens.

** As the sugar content in mirin is high, ensure that the flame is carefully watched over and is kept low.

** Adjust the sugar level as per personal taste when making the sauce.

** Sundried oysters are to be covered in full to soak in the mirin sauce. Refrigerate the remaining sauce.

** Besides using sundried oysters, this sauce can be used for marinating chicken wings, veal ribs and cods etc.

Chopped Dried Oysters in Lettuce Wrap

蠔豉鬆生菜包

InGreDienTs 材料

- 蠔豉 ....4 隻
  處理好，切碎
- 蝦仁 ....2 隻
  用鹽抓洗 2 次，沖水後索乾，切粒
- 馬蹄 ....4 個
  去皮，拍扁再剁碎
- 油條 ....1 條
  剪碎，冷油下鑊，加熱翻炸
- 青、紅椒 .... 各半隻
  去籽，切粒
- 西生菜 ....1 棵
  洗淨，摘出一片片
- 蒜頭 ....2 粒
- 海鮮醬 .... 隨量

- 4 dried oysters :
  treated and chopped
- 2 shelled shrimps : rubbed and squeezed with salt twice, washed and wiped off the moisture. Diced
- 4 water chestnuts :
  skinned, flattened and chopped
- 2 flour-stick : cut and put it in cold oil, turn on the heat and fried over medium heat.
- ½ each red and green peppers :
  seeds removed and diced
- 1 stalk lettuce :
  washed and torn into big pieces
- 2 cloves garlic
- some Hoisin sauce

MeThod 做法

❶用 2 湯匙油爆香蒜頭，放下蝦仁，爆炒至變紅色，盛起。

❷用餘下的油炒蠔豉、馬蹄、青紅椒約 1 分鐘，加入蝦仁及油條，炒勻，盛起，用生菜包蠔豉鬆，加海鮮醬伴食。

❶ Fry the garlic in 2 tbsps of oil till fragrant. Add the shrimps to fry till they turn red. Scoop.

❷ Fry the dried oysters, water chestnuts and red and green pepper in the remaining oil for about 1 minute. Add the shrimps and flour-stick. Stir well and scoop up. Wrap them with lettuce and seasoned with the Hoisin sauce to serve.

海味・淡菜 DRIED MUSSELS

挑選・SELECTING

購買淡菜時，宜選完整沒缺損、飽滿、顏色呈棕紅、略帶光澤的，同時如環境許可宜用手觸是否乾爽、嗅一嗅是否有霉味。貨品乾爽、沒霉味是購買海味的首要因素。

Select those which are not torn; brownish red in color and with sheen. If possible, touch and smell them whether they are dry and without the smelly odor. Being dry and without moldy smell are the top considerations for selecting dry goods.

處理・TREATING

淡菜用溫水浸軟，撕去淡菜的鬚，洗淨即可烹調。

Soak dried mussels in warm water till softened. Tear off the beard, wash them to use.

用途・USAGES

多用作湯料

Used mainly for making soups.

Dried Mussels and Bitter Melon Soup

淡菜涼瓜湯

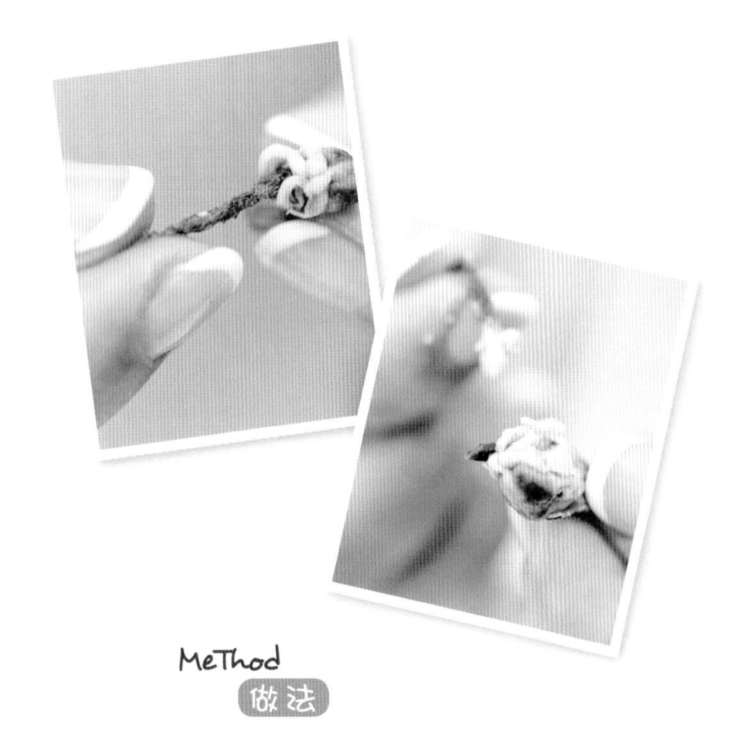

InGreDienTs 材料

- 淡菜 ....75 克
  處理好
- 涼瓜 ....2 個
  去瓤，切塊
- 眉豆 ....1/4 杯
  洗淨，用水略浸
- 黑豆 ....1/4 杯
  洗淨，用水略浸
- 赤小豆 ....1/4 杯
  洗淨，用水略浸
- 蜜棗 ....4 粒
- 陳皮 .... 半個
  浸軟，去瓤
- 水 ....10 杯

- 75 g dried mussels : treated
- 2 bitter melons : pith removed and cut into pieces
- ¼ cup black-eyed beans : washed and soaked briefly in water
- ¼ cup black beans : washed and soaked briefly in water
- ¼ cup small red beans : washed and soaked briefly in water
- 4 candied dates
- ½ dried tangerine peel : softened and pith removed
- 10 cups water

MeThod 做法

❶三色豆同放器皿內，加過面水置微波爐「叮」10 分鐘兩次，第二次加入淡菜同「叮」。
❷煲滾水後放入全部材料，用大火煲滾後再調中慢火煲半小時即成。

❶ Place the three beans into a container, cover them with water and put in the microwave oven to cook twice, each time for 10 minutes. At the second cooking, add the dried mussels.
❷ Bring the water to the boil, put all the ingredients in and boil over medium low heat for ½ hour. Serve.

Dried Mussels and Bitter Melon Soup

淡菜涼瓜湯

挑選 • SELECTING

宜挑選有海味鮮香、乾爽的螺頭，其中以澳洲出產的紅螺頭質素較佳。

Go for the fresh, dried and aromatic ones, of which the red dried conches from Australia are considered the best.

處理 • TREATING

用水浸半日至略軟，再用牙刷擦去螺頭的污漬，洗淨即可。

Soak them in water for half a day till they have softened slightly; use a toothbrush to brush off their impurities. Wash and they are ready for use.

用途 • USAGES

多用作煲湯或燉湯。

Mainly used for boiling and stewing soups.

InGreDienTs 材料

- 螺頭 ....225 克
  處理好
- 瘦肉 ....300 克
  汆水
- 金華火腿 ....2 塊
  切去肥肉，汆水
- 雞腳 ....4 對
  汆水
- 淮山 ....8 片
  沖洗後略浸
- 杞子 ....1 把
  沖洗後略浸
- 薑 ....2 片
- 水 ....12 杯

** 不要剪去雞腳的腳趾，否則雞腳會大量出油。

- 225 g dried conches : treated
- 300 g lean pork : scalded
- 2 pieces Jinhua ham :
  fats discarded and scalded
- 4 pairs chicken feet : scalded
- 8 slices Huai Shan :
  washed and soaked briefly
- 1 handful Qi Zi :
  washed and soaked briefly
- 2 slices ginger
- 12 cups water

MeThod 做法

❶所有材料放入水中，用大火煮滾，5 分鐘後改中小火再煲 3 小時即可。

❷飲湯前試味才下鹽(因螺頭及火腿均有鹹味)。

❶ Place all the ingredients in the water. Bring to the boil over high heat; after 5 minutes, turn the flame down to low and boil for 3 hours.

❷ Before serving, taste it to season with salt (as dried conches and ham are both salty)

** Do not cut off the claws of chicken feet as they would then ooze a lot of oil.

Dried Conches, Chicken Feet with Huai Shan and Qi Zi Soup

螺頭鳳爪淮山杞子湯

挑選 • SELECTING

大地魚以乾身、顏色金黃、有魚香味的為上品，宜購買大尾、連頭尾的。

The excellent ones are dry and flavorful. It is best to buy a big fish with head and tail intact.

處理 • TREATING

不要用水沖洗大地魚，用扭乾的濕毛巾抹乾淨即可；不要以為乾乾的大地魚很難處理，只要剪去魚頭和魚鰭，撕去皮，去骨，將肉剪成小方塊，用中火將魚塊炸至金黃色和香脆，盛起瀝去油分，即可使用。
將剩下的魚骨焗香，可用作煲湯的配料，緊記要放入煲湯魚袋內，避免飲湯時有魚骨而大煞風景。

Don't use water to wash da-ti fish. Twist dry a piece of wet cloth to wipe clean the fish. It is not difficult to clean it, all we need do is to cut off its head and fins, tear off the skin and remove the bones; slice them into small pieces and fry over medium heat till they are golden brown and crispy. Scoop up to drip off the oil for use.
The remaining fish bones when roasted well can be used as accompaniment for soups. Remember to place them in a fish bag when boiling soup to avoid having fish bones in the soup.

用途 • USAGES

湯底(雲吞麵)、用已舂碎的炸大地魚佐粥、燜菜、炒菜等等。

Used as soup base for wonton noodles. Chopped da-ti fish can also be added to porridge, and to be braised and stir fried.

手機一嘟即時去片！
http://e.formspub.com/videos/?id=dadi

Braised Da-Ti Fish with Pork Belly

大地魚燜五花腩

InGreDienTs 材料

- 大地魚肉 ....1 杯
- 五花腩 ....600 克
  切塊，出水
- 冬菇 ....5 朵
  處理法參閱 P.80
- 紹酒 ....2 湯匙
- 老抽 ....2 湯匙
- 魚露 ....1 湯匙

- 1 cup da-ti fish
- 600 g pork belly :
  cut and blanched
- 5 dried shiitake mushrooms :
  refer to pg: 80 for its treatment
- 2 tbsps Shaoxing wine
- 2 tbsps dark soy sauce
- 1 tbsp fish sauce

AccomPaniMents 配料

- 薑 ....2 片
- 蒜頭 ....2 粒
- 八角 ....1 粒
- 紅椒 ....1 隻
- 冰糖 ....1 粒

- 2 slices ginger
- 2 cloves garlic
- 1 star anise
- 1 red chilli
- 1 cube rock sugar

MeThod 做法

❶用 1 湯匙油爆香配料，待冰糖開始溶，倒下五花腩，兜爆一會，灒酒。
❷再加入大地魚及冬菇，下老抽添色，兜炒幾下，加滾水至肉面。
❸用中火燜 45 分鐘至汁稠，加魚露，兜勻即成。

❶ Fry the accompaniments with 1 tbsp of oil till fragrant. When rock sugar dissolves, pour the pork belly in and stir fry for a while. Sprinkle wine on side of wok.
❷ Add the fish and mushrooms. Sprinkle dark soy sauce on it, stir fry briefly. Add boiling water to cover the meat.
❸ Braise over medium heat for 45 minutes till the sauce thickens. Add the fish sauce and stir well before serving.

Chouzhou Da-ti Fish Porridge

大地魚潮州粥

Ingredients

材料

- 大地魚 ....1 條
  已處理
- 免治豬肉 ....150 克
  以半茶匙粟粉及少許胡椒粉略醃
- 蠔仔 ....600 克
  用粟粉和鹽抓洗兩次
- 冬菇 ....4 朵
  處理後，切絲
- 冷飯 ....3 碗
- 水 ....9 碗

- 1 da-ti fish : treated
- 150 g minced pork : marinated briefly with ½ tsp of cornflour and pinch of pepper
- 600 g oysters : rubbed and washed twice with cornflour
- 4 dried shiitake mushrooms : treated and shredded
- 3 bowls cooked rice
- 9 bowls water

Accompaniments

配料

- 炸香蒜片 .... 半杯
- 芫茜碎 .... 半杯
- 葱花 .... 半杯
- 大地魚碎 ....1 杯
- 魚露 .... 半杯
- 胡椒粉 .... 適量

- ½ cup deep fried sliced garlic
- ½ cup chopped coriander
- ½ cup chopped spring onion
- 1 cup da-ti fish powder
- ½ cup fish sauce
- pinch of pepper

Method

做法

❶將已炸香的大地魚放在攪拌機內打碎。
❷煲滾 9 杯水，即可加入冷飯，順序把冬菇絲、肉碎、蠔仔放入粥內，把材料攪開，再滾即成。
❸吃前可因應各人不同口味加入配料。

❶ Place the fried da-ti fish in a blender to grind into powder.
❷ Bring 9 cups of water to the boil. Add rice and then put the shredded mushrooms, minced pork and oysters (in this order) in the porridge, stir well. Scoop after the porridge is boiled.
❸ Before serving, add the accompaniments individually; use the amount as per personal preference.

＊＊ 正式的潮州粥有如廣東人的湯飯，所以米飯不可以煲至「開花」（糜爛）。
＊＊ 以上的材料是 6 人份量。

** Like the Cantonese soup and rice, the typical Chouzhou rice is not to be overcooked.
** This recipe is for 6 persons

海味・鹹魚 SALTED FISH

挑選 • SELECTING

優質的鹹魚有誘人的鹹香味、沒有蟲蛀、沒有霉味，肉質帶有光澤。如你是買整尾魚，一定要檢查魚頭及魚腩是否有發霉。

Good quality salted fish has the salted flavor, without worms, moldiness and the flesh has a nice sheen to it. If you are buying a whole fish, check whether the head and belly are moldy.

處理 • TREATING

原來用鹹魚做菜餚，並不是隨便取出所需份量就烹調，例如煲湯的鹹魚頭，就要先用油加薑塊煎香；炒飯及蒸肉餅的鹹魚肉要用油慢火炸香，盛起待冷後拆骨起肉，再撕成茸。

Adding salted fish to dishes take some proper handling. For example, salted fish head used for boiling soup would need to be fried till fragrant with oil and ginger slices; salted fish meat is to be deep fried until fragrant over low flame and scoop and let cool. Remove the bones and tear into pieces before using it for frying rice and steam with pork patty.

用途 • USAGES

鹹魚通常用來蒸或煎作菜餚，如惹味的鹹魚雞粒炒飯、鹹魚蒸肉餅等等，也有人用鹹魚頭煲豆腐湯，據說有下火的功效。

Used normally for steaming and frying, such as the salted fish with chicken fried rice and steamed salted fish with pork patty. Salted fish head is also used to boil in beancurd soup, which some say has the cooling effect on body.

Steamed Salted Fish and Grass Carp Fillet

生死戀

InGreDienTs 材料

- 梅香鹹魚 ....1 塊
  約 150 克，處理好
- 鯇魚腩 ....1 塊
  約 200 克
- 薑 ....4 大片
  去皮，切絲

- 1 piece salted fish :
  about 150 g, treated
- 1 piece grass carp fillet :
  about 200 g
- 4 big slices ginger :
  skinned and shredded

MeThod 做法

❶碟內放鯇魚腩，放上鹹魚茸，鋪上薑絲，隔水蒸 8 分鐘。

❷灒上 1 湯匙熟油及 1 湯匙生抽即成。

＊＊ 生死戀，鮮魚腩是生，鹹魚則已死，取其悽美之意。

＊＊ 可將魚放入微波爐「叮」3 分鐘，代替隔水蒸。

❶ Place the grass carp fillet on a plate. Put the salted fish on top of the fillet. Spread the shredded ginger over them. Steam over water for 8 minutes.

❷ Pour 1 tbsp of hot oil and 1 tbsp of light soy sauce on the fish. Serve.

** Alternatively, cook the fish in microwave oven for 3 minutes instead of steaming it over water.

Steamed Salted Fish and Grass Carp Fillet

生死戀

海味 • 乾魷魚 DRIED SQUID

挑選 • SELECTING

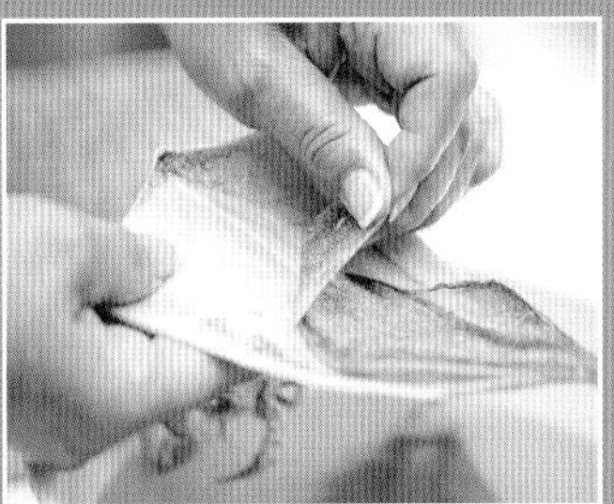

宜選肉質透微紅、微帶些白粉（這些白色的粉末是海水結晶）、乾身、嗅時有濃濃魷魚香的乾魷魚。可能會因製作技術和貯藏欠佳，乾魷魚會容易蟲蛀或有霉味，購買時要留意。

Pick those whose flesh is reddish and covered with whitish powder (the crystallized sea salt), they are dry and have strong squid's smell. Due to the bad treatment and storage, dried squid can get to be worm infested and moldy.

處理 • TREATING

乾魷魚用溫水浸軟，撕去薄膜及軟骨就可採用。若想炒魷魚時呈漂亮的花紋，在有軟骨的一邊斜刀切菱形花紋即可。

Soak dried squid in warm water till they are soft. Tear off the thin membrane and soft bones for use. To show the pretty diagonal lines on squid, use a knife to cut crisscross lines on the sides of the soft bones.

用途 • USAGES

蒸、炒、燜皆宜。

Suitable for steaming, stir frying or stewing.

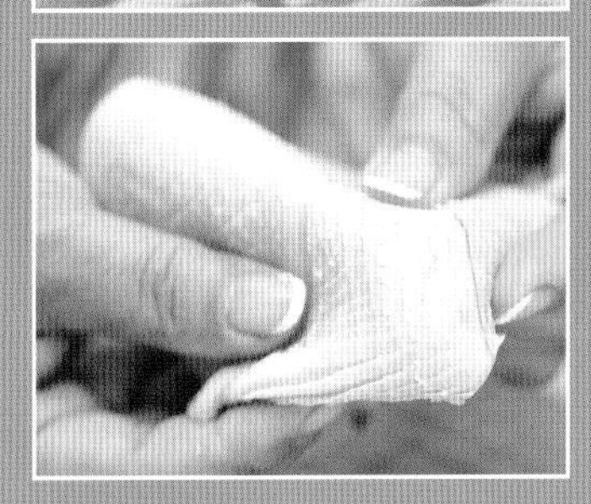

Dried Squid and Pork Balls

魷魚粒豬肉丸

InGreDienTs
材料

- 免治豬肉 ....300 克
  加醃料略醃
- 乾魷魚 ....1 條
  處理後，切碎
- 馬蹄 ....3 粒
  去皮，拍扁，剁碎
- 芹菜 ....1 棵
  去葉，切粒
- 水 ....4 杯

- 300 g minced pork : marinated briefly with the marinade
- 1 piece dried squid : treated and chopped
- 3 water chestnuts : skinned, flattened and chopped
- 1 stalk celery : leaves discarded and diced
- 4 cups water

MariNade
醃料

- 鹽 ....1/2 茶匙
- 糖 ....1/4 茶匙
- 胡椒粉、麻油 .... 各少許
- 粟粉 ....1 湯匙

- 1/2 tsp salt
- 1/2 tsp sugar
- dash of pepper and sesame oil
- 1 tbsp cornflour

MeThod
做法

❶將魷魚粒、馬蹄粒和免治豬肉拌勻，撻撻至起膠，再擠成肉丸。
❷煲滾水，放入肉丸，煲至肉丸浮起，灑芹菜粒即可享用。
＊＊芹菜香氣濃郁，令這湯菜生色不少。
＊＊馬蹄的嫩芽部位和底部藏了不少污泥，所以要先切去才清洗馬蹄及去皮。

❶ Mix the chopped squid, chopped water chestnuts and minced pork well, rub and beat till sticky to the touch. Rub into balls.
❷ Bring a pot of water to the boil, add the meat balls. When the balls are afloat, sprinkle with diced celery to serve.
** The aroma of celery adds more flavor to the soup.
** The buds of water chestnut often are imbedded with dirt, cut it off before washing them and removing skins.

海味・蝦乾 DRIED SHRIMPS

挑選 • SELECTING

購買蝦乾時，除了用手觸它是否乾爽外，也要嗅一嗅它是否有鮮香的蝦味，如有陣陣霉味就不要買了。

Touch them with hands to feel whether they are dry; put them to the nose as well. If there is some moldy smell, they are not fresh.

處理 • TREATING

蝦乾沖洗乾淨後，用溫水略浸 8 分鐘取出，加糖、生抽拌勻待用。用溫水浸蝦乾可加速軟化，並可保留原味。

After washing and cleaning them, Soak briefly in warm water till softened. Remove and season with sugar and light soy sauce to mix well. Soaking them in warm water speeds up the softening process and their flavor is kept well.

用途 • USAGES

蒸、炒皆宜。

Suitable for steaming or stir frying.

Dried Shrimps and Dried Squid Rice

蝦乾魷魚飯

Ingredients 材料

- 魷魚 ....1 條
  處理好，切粒
- 蝦乾 ....75 克
  處理好
- 菜脯 ....10 條
  略浸，切粒，用白鑊烘乾，加 1 茶匙油、1 茶匙糖，關火，拌勻，盛起待用
- 芋頭 ....1/4 個，約 300 克
  去皮，切粒，炸脆
- 葱 ....2 棵 切粒
- 生抽 .... 適量
- 米 ....3 杯
- 水 ....2 1/2-3 杯

- 1 dried squid : treated and diced
- 75 g dried shrimps : treated
- 10 pieces preserved radish : Soaked briefly and diced. Stir fry in a dry wok to reduce its water content over low heat. Add 1 tbsp of oil and 1 tsp of sugar. Remove from heat. Mix well.
- 1/4 taro : about 300 g, skinned, diced and deep fried till crispy
- 2 sprigs spring onion : diced
- some light soy sauce
- 3 cups rice
- 2 1/2-3 cups water

Method 做法

❶米淘洗乾淨，放入電飯煲內，注入水，按掣，待煲飯水成蝦眼水，放入所有材料（除芋頭、生抽、葱外）。

❷待電飯煲跳掣後，倒入芋頭、生抽拌勻，再次按電掣，待再次跳掣後，倒入葱粒，撈勻即可享用。

＊＊ 米有新、舊之分，各有不用的吸水性，舊米須用 3 杯水，而新米用 2 1/2 杯就可以了。

❶ After washing the rice, place in a rice cooker, add the water to switch it on to cook. Till the first boiling bubbles appear, place all the ingredients (except diced taro, light soy sauce and spring onion) in the pot.

❷ When it is switched off, pour the diced taro in and mix well with the light soy sauce. Press the switch again and when it gets switched off the second time, put in the diced spring onion. Stir well to serve.

** Rice has the "old" and "new" categories. Use 3 cups of water for boiling "old" rice but use only 2 ½ cups water for the latter.

Dried Shrimps and Dried Squid Rice

蝦乾魷魚飯

海味・銀魚乾
SMALL DRIED FISH

挑選 • SELECTING

宜選購乾燥、呈淡黃色的。

Choose the dry and light yellow colored dried fish.

處理 • TREATING

將銀魚乾用水沖洗後，換上溫水浸至軟身約 10 分鐘，即可瀝乾水分，用薑汁酒、生抽拌勻待用。

Wash dried fish in water. Afterwards, change it to soak in warm water till they soften; drip dry.
Use ginger wine and light soy sauce to season.

用途 • USAGES

銀魚乾味道鮮香，用來蒸、炒皆宜。

Small dried fish is tasty and good for both steaming and stir-frying.

Steamed Dried Fish with Minced Pork

銀魚乾蒸豬肉餅

InGreDienTs 材料

- 免治豬肉 ....300 克
- 銀魚乾 ....30 克
  處理好
- 水 ....4 湯匙
- 薑 ....2 片
  切絲

- 300 g minced pork
- 30 g small dried fish : treated
- 4 tbsps water
- 2 slices ginger : shredded

SeaSonIng 調味料

- 鹽 ....1/2 茶匙
- 糖 ....1/4 茶匙
- 粟粉 ....2 茶匙
- 胡椒粉 .... 少許

- ½ tsp salt
- 1/4 tsp sugar
- 2 tsps cornflour
- pinch of pepper

MeThod 做法

❶免治豬肉和調味料拌勻，逐少加入水（期間要不斷攪拌，待肉吸收水分才再加），鋪上銀魚乾和薑絲，隔熱水猛火蒸 12 分鐘。

❶ Season the minced pork well. Add water a little at a time. Stir and mix continually; rub water into pork only when it is fully absorbed. Spread the small dried fish and shredded ginger over it. Steam over hot water and high heat for 12 minutes.

海味•海蜇 SALTED JELLYFISH

挑選 • SELECTING

店舖內有整片用鹽醃及已浸泡、去鹹味的海蜇絲出售，悉聽尊便。

Available for sale in shops are salted jellyfish in whole piece, or pre-soaked and unsalted shredded jellyfish .

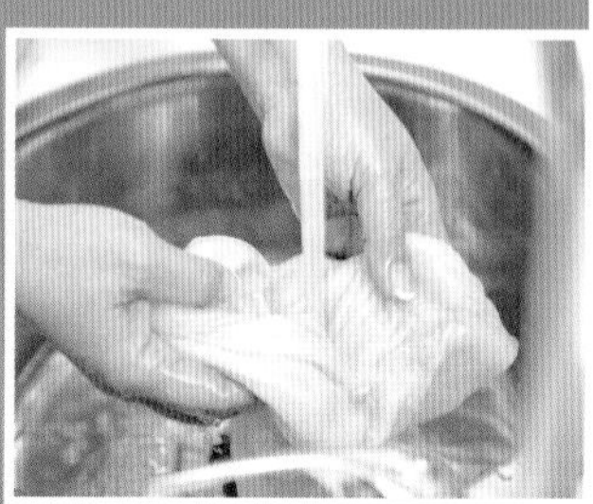

處理 • TREATING

如買整片鹽醃海蜇，須將海蜇浸一夜，洗去鹽分，試味（如仍覺太鹹，須再浸泡），切絲；海蜇絲倒入筲箕內，沖入大滾熱水，瀝乾，即可採用。
如買已浸泡、去鹹味的海蜇絲，都要試味；如味道適中，可用凍滾水沖淨，瀝乾，就可調味了。

Soak the whole piece of salted jellyfish overnight to rinse off the saltiness. Soak it for a longer time if it is still too salty. Shred the jellyfish and place it onto a sieve, pour hot water over it and drip dry for use. However, if the shredded jellyfish is pre-soaked and unsalted, sample it first. If the taste is acceptable, rinse it with cold water and drip dry before seasoning.

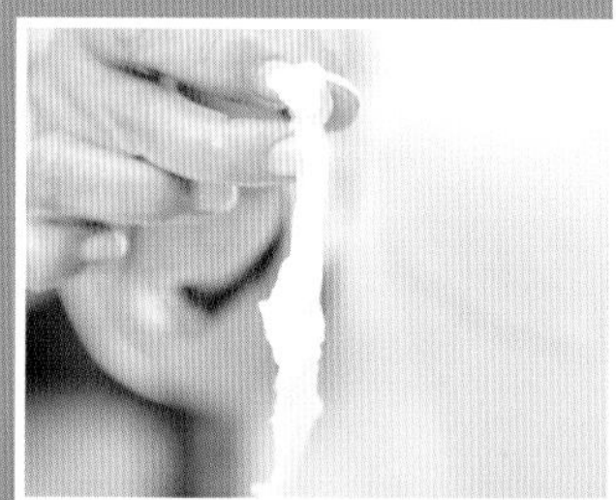

用途 • USAGES

煲湯、涼拌。

Use it for boiling soup and in cold dish.

Mixed Salted Jellyfish and Cucumber Salad

海蜇拌青瓜

InGreDienTs 材料

- 海蜇 ....100 克
  處理好
- 小青瓜 ....4 條約 300 克
  切去頭尾並用來磨切口，
  待磨出黏液，洗淨
- 紅椒 ....1 隻
  斜刀切塊，去籽

- 100 g salted jellyfish : treated
- 4 about 300 g cucumbers :
  Cut the two ends and use them to rub on the cucumber. Washed.
- 1 red chili : sliced diagonally and seeds removed

SeaSonIng 調味料

- 蒜茸 ....1 湯匙
- 生抽 ....1 茶匙
- 鎮江香醋 ....1 湯匙
- 糖 ....1 湯匙
- 麻油 .... 少許

- 1 tbsp minced garlic
- 1 tsp light soy sauce
- 1 tbsp Zhenjiang black vinegar
- 1 tbsp sugar
- dash of sesame oil

MeThod 做法

❶用刀拍扁青瓜，再加幾刀切成度，加 1 茶匙鹽醃 1 小時後，滗去水，索乾待用。
❷海蜇亦擠乾水分，連同青瓜及紅椒加入調味料醃即成。
＊＊ 用青瓜頭尾磨青瓜切口，可去苦味。
＊＊ 海蜇和青瓜最重要的是要醃得入味、均勻，各位可將海蜇、青瓜和調味料一起放在保鮮袋內搖勻，就可享用了。

❶ Flatten the cucumber with the back of a knife, and then section it. Add 1 tsp of salt to marinate for 1 hour. Discard the water and squeeze dry for later use.
❷ Squeeze dry the jellyfish, marinate together with the cucumber and red chili. Season to serve.
** Rub cucumbers with its cut head and tail eliminates its bitter taste.
** Ensure that salted jellyfish and cucumbers are marinated thoroughly. Leave them in a plastic bag and shake vigorously to mix well.

海味・蝦子
DRIED SHRIMP ROE

挑選 • SELECTING

蝦子是蝦的乾燥卵子，以乾身、顆粒圓、味淡、顏色紅或金黃色為佳。

Good dried shrimp roe should be dry, round in shape, light in taste and red or golden yellow color.

處理 • TREATING

蝦子要辟去腥味，才可嘗到鮮味，有說要在蝦子內加點紹興酒、兩片薑，隔水蒸 15 分鐘，取出，挑走薑片，再放在白鑊烘乾，但要用小火及要有耐性，否則蝦子焦了便會變苦。日前，席間遇鏞記甘健成先生，他教我一個更容易的方法，用 2 粒蒜頭和蝦子一起用白鑊慢火炒香即成。

First, get rid of its fishy smell in order to taste its freshness. Add some Shaoxing wine, together with two slices of ginger to mix in them. Steam them over water for 15 minutes. Remove the ginger slices and leave in a dry wok to reduce its water content over low heat. Be patient or they taste bitter if burnt.
I met Mr. Kinsen Kam of Yung Kee Restaurant recently and he taught to me an easier way -fry dried shrimp roe in a dry wok with 2 cloves of garlic over low heat till fragrant!

用途 • USAGES

蝦子是增香、提鮮的綠葉角色，有它令菜式生色不少，如蝦子柚皮、齋鵝等等。

It is a flavor enhancer to dishes and used in such dishes as dried shrimp roe with dried pomelo skins and vegetarian goose etc.

Vegetarian Goose

蝦子齋鵝

InGreDienTs 材料

- 腐皮 ....4 張
- 五香粉 .... 半茶匙

- 4 dried beancurd sheets
- ½ tsp five spice powder

SeaSonIng 調味料

- 雞湯 ....3 杯
- 糖 ....2 茶匙
- 老抽 ....2 茶匙
- 麻油 ....1 湯匙
- 胡椒粉 .... 少許
- 蝦子 .... 隨意

- 3 cups chicken stock
- 2 tsps sugar
- 2 tsps dark soy sauce
- 1 tbsp sesame oil
- pinch of pepper
- some dried shrimp roe

MeThod 做法

❶將調味料調勻，慢火煮滾後分成2份，其中一份加入五香粉成醃料。
❷用加了五香粉的醃料抹勻每張腐皮，一張疊一張，再將其覆摺成鵝胸形，用牙籤固定形狀。
❸熱鑊下油1湯匙，煎香素鵝兩面，倒入其餘調味料，改慢火，燜約15分鐘至汁稠，冷卻後取出切塊享用。

❶ Mix the seasoning well. Bring to the boil over low heat. Divide the seasoning ingredient into 2 portions, add the five spice powder to one of which as marinade.
❷ Spread the marinade on each beancurd sheet evenly, stack one atop another. Fold them into the shape of a goose breast, affix it with a toothpick.
❸ Fry both sides of the 'goose breast' in 1 tbsp of oil till fragrant, pour the rest of the seasoning in. Turn down to low heat to cook for 15 minutes till the sauce thickens. When cool, slice them to serve.

乾貨

如果說，海味是天生的主角，哪乾貨豈不是永恒的「茄喱啡」？

這是謬見。

只要處理、配搭得宜，乾草菇、雪耳、杞子、冬菇、合桃等等，也可發光發熱，成為整道菜的聚焦點。

Dried Goods

With the right methods, white fungus, Qi Zi, dried shiitake mushrooms, walnuts and many other dried goods, they can shine as main dishes as well!

挑選 • SELECTING

冬菇以肉厚、香味濃郁、花紋明顯、底部顏色淺、乾爽為佳。

The good ones are those whose flesh is thick and fragrant, its stripes are clear; their stems show a light color and are dry.

處理 • TREATING

如冬菇用作煲湯，將冬菇浸軟，剪去木屑和菇蒂，洗淨即可採用。如冬菇用作燜煮，將已浸軟洗淨的冬菇放入碗內，倒入已過濾的浸菇水（要浸過菇面，如水不足，須加添熱水），下 2 片薑、葱 2 棵和 1/4 片片糖，用保鮮紙封好，再刺兩個小孔，放入微波爐「叮」10 分鐘兩次。
用油爆香 2 粒蒜頭，下冬菇，炒勻，加入草菇老抽 2 湯匙，倒入浸菇水，煮至水分收乾，下蠔油 2 湯匙炒勻，關火，待涼。
將冬菇分數份裝在保鮮盒或保鮮膠袋內，放入冰格冷藏；烹調前不用解凍，可直接放入鍋內煮。

For boiling soup, soak dried shiitake mushrooms in water till softened, cut off the stems and woody bits. Wash them to be used. If it is for braising, place the already softened dried mushrooms in a bowl, pour the sifted mushroom water in (to cover the surface). Add some hot water if it is insufficient, add two slices of ginger, 2 stalks of spring onion and 1/4 slice of slab sugar. Wrap tightly with cling film, poke two holes on top and place in microwave oven to cook for 10 minutes, twice.
Fry 2 cloves of garlic till fragrant in oil, add the mushrooms to stir well. Add 2 tbsps of the mushroom-flavored dark soy sauce; pour the mushroom water in to cook till the water evaporates. Season with 2 tbsps of oyster sauce. Stir well. Switch off the flame and let it cool.
Divide the mushrooms into equal portions to keep in plastic boxes or bags. Freeze them. Do not defrost before using it for cooking; put it directly into the pot.

用途 • USAGES

煲湯、炒、燜、煮等。

Good for boiling soup, stir frying, braising and cooking.

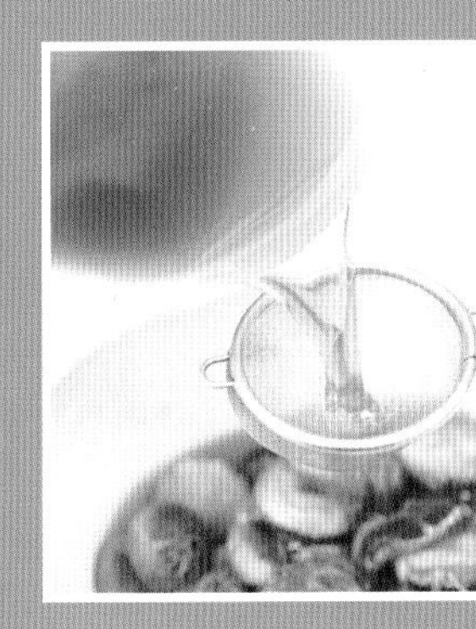

Stuffed Dried Shiitake Mushrooms

釀冬菇

InGreDienTs
材料

- 冬菇 ....12 朵
  處理好，索乾

- 12 dried shiitake mushrooms : treated and squeezed dry

Filling
餡料

- 蓮藕 ....1 小節約 300 克
  去皮，磨茸，漟去水，留茸
- 臘腸 ....1 條
  剁成茸
- 瑤柱 ....2 湯匙
  處理好，撕開，剁碎
- 芫茜 ....1 棵
  切至極碎

- 1 section lotus root about 300 g : skinned and minced; squeezed dry
- 1 dried pork sausage : minced
- 2 tbsps dried scallop : treated, shredded and chopped
- 1 stalk coriander : chopped finely

MariNade
醃料

- 澄麵 ....1 湯匙
- 鹽 .... 半茶匙
- 胡椒粉 .... 少許

- 1 tbsp Tang flour (wheat starch)
- 1/2 tsp salt
- pinch of pepper

MeThod
做法

❶在已索乾水分的冬菇蒂部抹少許生粉。
❷餡料與醃料混合後，釀入冬菇內，以中慢火煎香即成。
＊＊ 臘腸須略蒸才切片、剁茸，腸衣才不會在切時甩掉。

❶ Rub a little cornflour on the stems of the squeezed dry mushrooms.
❷ Mix the filling with the marinade to stuff inside the dried mushrooms. Fry over medium low heat to serve.
** Steam sausages briefly before slicing and mincing them to avoid the skin being peeled off.

Gong Xi Fa Cai

恭喜發財

InGreDienTs 材料

- 髮菜 ....8 克
  浸軟，去污，換過水後加 2 滴油蒸 10 分鐘；或「叮」3 分鐘
- 菠菜 ....450 克
  洗淨，摘斷
- 冬菇 ....8 朵 處理好
- 蒜頭 ....2 粒

- 8 g dried black moss : softened in water, impurities discarded. Change the water and add 2 drops of oil. Steam for 10 minutes or place in microwave oven to cook for 3 minutes.
- 450 g spinach : washed and broken in pieces
- 8 dried shiitake mushrooms : treated
- 2 cloves garlic

SeaSonIng 調味料

- 生抽 ....1 湯匙
- 蠔油 ....2 湯匙
- 麻油 .... 少許
- 糖 .... 半茶匙
- 浸冬菇水 .... 半杯

- 1 tbsp light soy sauce
- 2 tbsps oyster sauce
- dash of sesame oil
- 1/2 tsp sugar
- 1/2 cup water used for soaking the dried mushrooms

MeThod 做法

❶煲滾 2 杯水，下 1 湯匙油、1 茶匙鹽，放入菠菜焯 1 1/2 分鐘，瀝乾水分，置碟中。
❷用 1 湯匙油爆香蒜頭和冬菇，下髮菜和調味料燜 15 分鐘。
❸將冬菇圍着菠菜，髮菜堆在菠菜上，舀上燜汁即可享用。
＊＊ 焯菜只需注入 2 吋水，不相信？就試試吧！

❶ Bring 2 cups of water to the boil; add 1 tbsp of oil and 1 tsp of salt. Put the spinach in to blanch for 1 1/2 minutes. Drip dry and place on a plate.
❷ Fry the garlic and dried mushrooms in 1 tbsp of oil till fragrant. Add the black moss and seasoning to braise for 15 minutes.
❸ Surround the spinach with the dried mushrooms; place the black moss on top of the spinach. Pour the sauce over it to serve.
** Use only 2 inches of water to blanch vegetables, instead of filling the pot half full. Try it to believe it!

Stewed Dried Mushrooms with Free Range Chicken

香菇燉走地雞

InGreDienTs 材料

- 冬菇 ....100 克
  處理好
- 走地雞 ....1 隻約 1.8 公斤
  劏乾淨，汆水
- 豬手尖 ....1 隻約 200 克
  洗淨，去毛，汆水
- 滾水 ....4-6 杯
- 薑 ....4 片

- 100 g dried shiitake mushrooms : treated
- 1 about 1.8 kg free range chicken : gutted and washed, scalded
- 1 about 200 g pig's trotter : washed, hair removed and scalded
- 4-6 cup boiling water
- 4 slices ginger

MeThod 做法

❶將所有材料放入燉盅，加水至八分滿，用沙紙封口，隔水燉 3 小時即成。

❶ Place all the ingredients into a stewing pot; add water till it is about 80% full. Seal the opening with mulberry paper. Stew over the water for 3 hours. Ready to serve.

Braised Deer Antler with Pig's Trotter

鹿茸燜豬手

InGreDienTs 材料

- 鹿茸 ....10 克
- 豬手 ....1 隻約 300 克
  斬件，汆水 15 分鐘
- 冬菇 ....6 朵
  處理好
- 薑 ....2 片
- 葱 ....2 條

- 10 g deer antler
- 1 about 300 g pig's trotter : cut into pieces and scalded for 15 minutes
- 6 dried shiitake mushrooms : treated
- 2 slices ginger
- 2 stalk spring onion

SeaSonIng 調味料

- 蠔油 ....2 湯匙
- 草菇老抽 ....1 湯匙
- 鹽 ....1 茶匙

- 2 tbsps oyster sauce
- 1 tbsp mushroom-flavored dark soy sauce
- 1 tsp salt

MeThod

做法

❶用 2 湯匙油爆香薑片、豬手，倒入 2 杯滾水，再加冬菇、鹿茸、葱條及調味料，用小火燜 1 3/4 小時即成。

＊＊ 在不同品牌的老抽中，我愛用草菇老抽，因它最易令食材上色。

❶ Fry the ginger slices and pig's trotter in 2 tbsps of oil till fragrant. Add 2 cups of boiling water, and then add the mushrooms, deer antler, spring onion and seasoning. Braise over low heat for 1 3/4 hours. Serve.

** I prefer to season with mushroom-flavored dark soy sauce as it enhances the food color.

乾貨・乾草菇
DRIED STRAW MUSHROOMS

挑選・SELECTING

宜挑選乾爽、清香沒有霉味的乾草菇，如顏色呈深黃色就不要買了，因放置時間過久，香味不濃。

Pick the ones which are dry, fragrant and without any moldy odor. Don't choose those whose colors are dark yellow, as they are on the shelves so long that they are not fragrant.

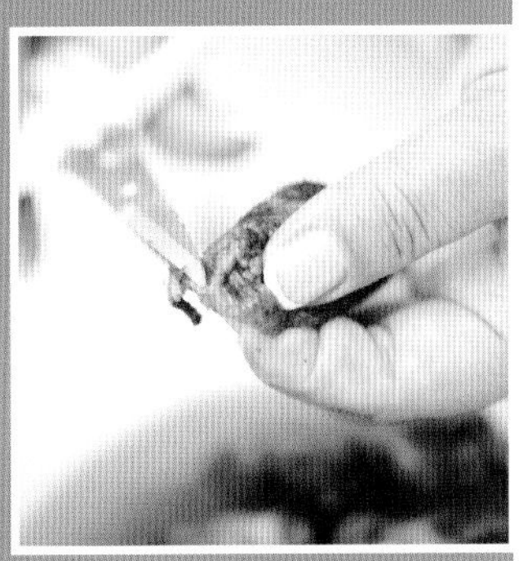

處理・TREATING

將乾草菇浸洗至軟身，剪去硬蒂後，用牙刷擦去泥污，洗淨，待用。用密筲箕隔去浸草菇水的污物，水留用。

Soak the dried straw mushrooms in water till they are softened. Cut off the hard stems. Use a toothbrush to clean off the surface dirt. Wash and set aside. Sift the water to discard the impurities, reserve it for use.

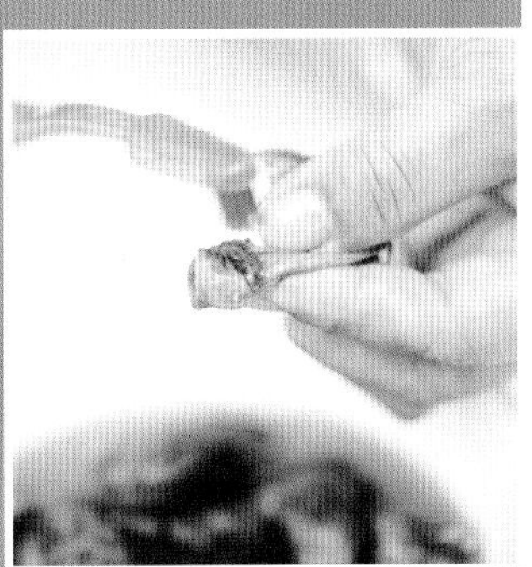

用途・USAGES

乾草菇用途廣泛，可用作煲湯、蒸、炒、煮、燜；因為浸過乾草菇的水特別香，故一般留作烹調用。

Used in boiling soup, steaming, stir frying and braising. The water used for soaking dried mushrooms is fragrant hence it is used to cook with the mushrooms.

Fresh and Dried Straw Mushrooms with Zhu Sheng Soup

鮮陳草菇竹笙湯

InGreDienTs
材料

- 乾草菇 ....38 克
  處理好
- 鮮草菇 ....150 克
  用扭乾濕布抹乾淨草菇，
  然後放入薑葱滾水內飛水
- 竹笙 ....19 克
  參閱第 112 頁處理法，再切段
- 浸乾草菇水 ....4 杯
- 水 ....3 杯

- 38 g dried straw mushrooms : treated
- 150 g fresh straw mushrooms : cleaned with a damp cloth, then scalded in boiling water with spring onion and ginger
- 19 g Zhu Sheng : refer to page 112 for its treatment and sectioned
- 4 cups water used for soaking the dried straw mushrooms
- 3 cups water

MeThod
做法

❶煲滾乾草菇水，放入乾草菇、竹笙，滾起後放鮮草菇，再滾 1 分鐘即可享用。

❶ Bring 3 cups of the dried mushroom soaking water to the boil; add the dried straw mushrooms and Zhu Sheng. When it is boiling, add fresh mushrooms, boil for another 1 minute. Serve.

Four Mushrooms Stew

乾草菇燴三菇

InGreDienTs 材料

- 乾草菇 ....50 克
  處理好，浸草菇水留用
- 鮮冬菇 ....100 克
  用扭乾濕布抹乾淨
- 秀珍菇 ....100 克
  用扭乾濕布抹乾淨
- 金針菇 .... 一包
  剪去尾部
- 豆腐 ....1 塊
  汆水後切塊
- 蒜頭 ....2 粒
- 紹酒 ....1 湯匙
- 葱 ....2 條
  切碎成葱花

- 50 g dried straw mushrooms:
  treated and reserve the water for use
- 100 g shiitake mushrooms :
  cleaned with a damp cloth
- 100 g fresh abalone mushrooms :
  cleaned with a damp cloth
- 1 packet enokitake mushroom :
  cut off the bottom impurities
- 1 cube beancurd :
  blanched in water and sectioned
- 2 cloves garlic
- 1 tbsp Shaoxing wine
- 2 stalks spring onion : chopped

SeaSonIng 調味料

- 生抽 ....1 湯匙
- 蠔油 ....1 湯匙
- 草菇水 .... 半杯
- 糖 ....1 茶匙
- 胡椒粉 .... 少許
- 麻油 .... 少許

- 1 tbsp light soy sauce
- 1 tbsp oyster sauce
- 1/2 cup water used for soaking the dried straw mushrooms
- 1 tsp sugar
- pinch of pepper
- dash of sesame oil

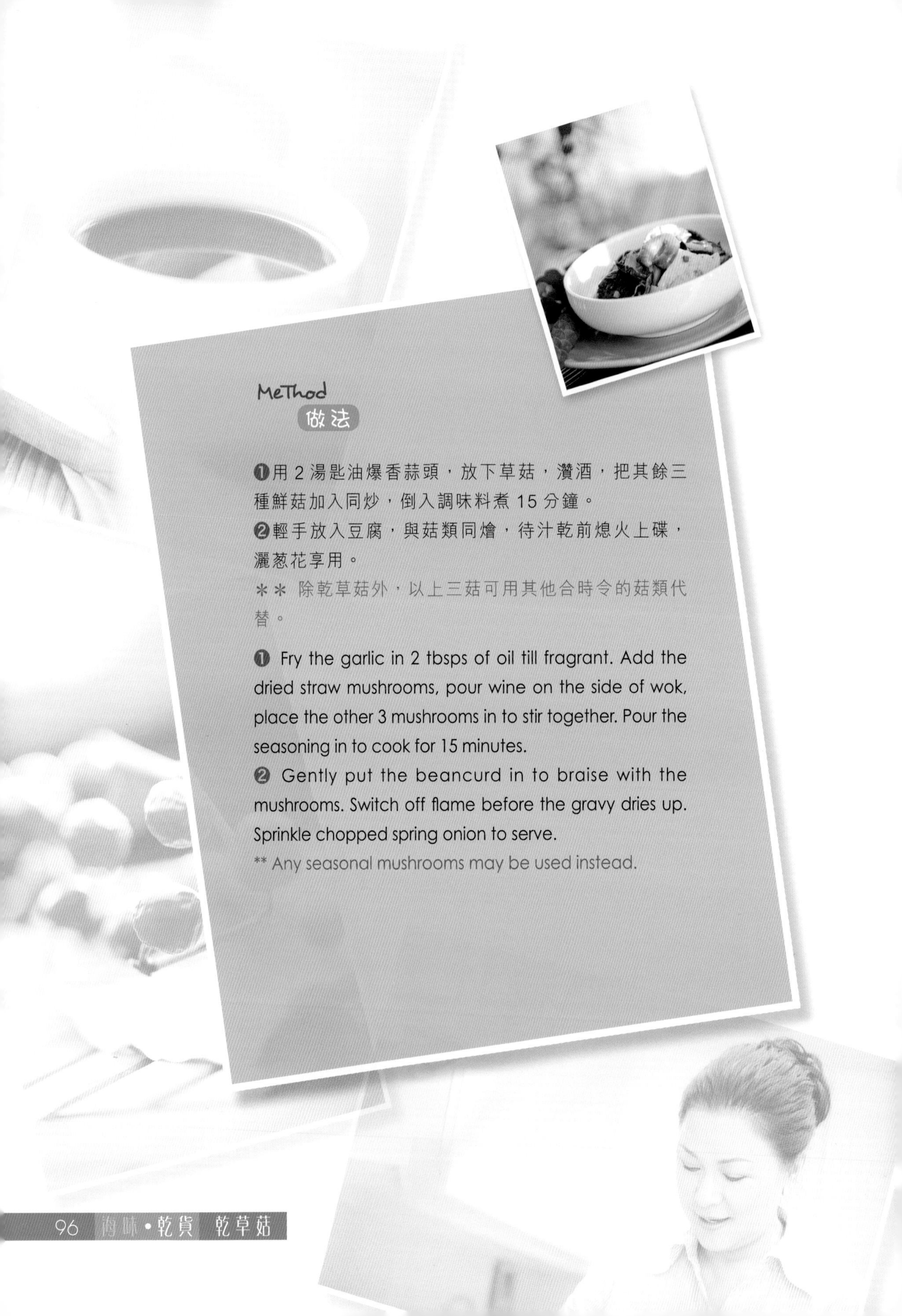

MeThod
做法

❶用 2 湯匙油爆香蒜頭，放下草菇，灒酒，把其餘三種鮮菇加入同炒，倒入調味料煮 15 分鐘。
❷輕手放入豆腐，與菇類同燴，待汁乾前熄火上碟，灑葱花享用。
＊＊ 除乾草菇外，以上三菇可用其他合時令的菇類代替。

❶ Fry the garlic in 2 tbsps of oil till fragrant. Add the dried straw mushrooms, pour wine on the side of wok, place the other 3 mushrooms in to stir together. Pour the seasoning in to cook for 15 minutes.
❷ Gently put the beancurd in to braise with the mushrooms. Switch off flame before the gravy dries up. Sprinkle chopped spring onion to serve.
** Any seasonal mushrooms may be used instead.

挑選 • SELECTING

凡購菇菌類的乾品，首要的條件是乾燥，雪耳也不例外。相信你們也知道不要買顏色潔白或過黃的木耳，宜買淡黃色的，如還是猶疑，聞一聞是否有刺鼻的漂白水或硫磺味，如有，就不要買了。

The first criterion for selecting mushrooms and fungus is their dryness. White fungus is no exception. You may already know not to choose the bleached white or too yellow looking fungus. When in doubt, smell whether the fungus has bleaching or sulphur odor.

處理 • TREATING

雪耳用清水浸泡 3-4 小時，直至變軟發大，剪去蒂部的木屑，清洗乾淨即可使用。

Soak white fungus in water for 3-4 hours till they are softened and expanded. Cut off the woody bits at stems; wash and clean for use.

用途 • USAGES

雪耳本身沒有味道，但它有養陰補肺、止血通便、潤膚的功效，所以許多湯品菜式都見到它的踪影，如老火湯、燉湯、羹、甜品、燜菜等。

On its own, it is tasteless. However, it helps in improving our skin and strengthening the lungs, which explains why it is used in many soup dishes such as long-boiled soups, stewed soups, thick soups, desserts and braised food.

Colorful Porridge with White Fungus

銀耳五彩粥

- 日本米 .... 半杯
  用水浸過夜
- 水 ....8 杯
- 雪耳（銀耳）........1 湯匙
  處理好，撕成小朵
- 粟米粒 ....1 杯
- 急凍雜豆 ....2 湯匙
- 冬菇 ....2 朵
  參閱第 80 頁處理法，再切細粒
- 白豆腐乾 ....1 塊 切粒

- 1/2 cup Japanese rice :
  soaked overnight
- 8 cups water
- 1 tbsp white fungi :
  treated and torn into pieces
- 1 cup sweet corn
- 2 tbsps defrosted green peas & diced carrot
- 2 dried shiitake mushrooms :
  refer to page 80 for its treatment
  and cubed
- 1 dried beancurd : diced

MeThod 做法

❶將 8 杯水煲滾後，加入米煲 1 小時。
❷將其餘材料汆水 3 分鐘後加入粥內，攪勻，加鹽調味即可。

❶ Bring the 8 cups of water to the boil. Add rice and cook for 1 hour.
❷ Blanch all the other ingredients for 3 minutes. Put into the porridge. Stir it and season with salt to serve.

White Fungus and Minced Chicken Thick Soup

雪耳雞茸羹

InGreDienTs 材料

- 雞肉 ....150 克
  剁碎，以半茶匙鹽、少許胡椒粉和 2 湯匙水略醃
- 雪耳 ....38 克
  處理好，汆水
- 蛋白 ....2 個 打勻
- 水 ....8 杯

- 150 g chicken :
  minced and marinated with 1/2 tsp of salt, pinch of pepper and 2 tbsps of water briefly
- 38 g white fungi :
  treated and blanched
- 2 egg whites : beaten well
- 8 cups water

Thickening 薄獻

- 馬蹄粉 ....2 湯匙
- 水 ....2 湯匙

** 調勻

- 2 tbsps water chestnut starch
- 2 tbsps water

**Mix well

MeThod 做法

❶煮滾水後放入雪耳滾 3 分鐘，下雞茸 (邊放邊攪)，待雞茸變白色後，逐少加入薄獻（邊放邊攪），再把蛋白注入，攪拌均勻即可熄火，隨意加添胡椒粉調味。

❶ Bring the water to the boil. Add the white fungi to boil for 3 minutes, and then put the minced chicken, stirring while adding it. When the minced chicken turns whitish, pour the thickening, stirring while adding it. Stir in egg whites then switch off the flame. Sprinkle with pepper as you like.

乾貨・木耳 BLACK FUNGUS

挑選 • SELECTING

宜挑選面呈烏黑、背部灰白色、體輕、乾爽，沒有霉味、有淡淡菇香的木耳。

It is good to buy those which are darkish black on surface and grayish white on their back - light, dry, no moldy odor but with light fragrance.

處理 • TREATING

用清水浸泡至軟及膨脹至數倍(起碼需 3、4 小時)，剪去蒂部的木屑，清洗乾淨即可採用。

Soak black fungus in water till softened and expanded many times (at least 3-4 hours). Cut off the woody bits at stems; wash and clean them for use.

用途 • USAGES

可涼拌、燜、炒、煲湯及煲茶水。木耳一向被視作女性補品，有止血的功效，亦可滋補強壯、通便治痔、去瘀生新。

Used in mixing cold dishes, braising, stir frying, boiling soup and tea. It is often regarded as a female elixir to stop bleeding and to nourish and strengthen the body. It is helpful and promotes health for our digestive system.

Cold Mixed Double Fungi

涼拌雙耳

InGreDienTs 材料

- 木耳 ....50 克
  處理好，切粗條
- 銀耳 ....50 克
  參閱第 97 頁處理法，
  再剪成小朵
- 杞子 ....1 湯匙
  洗淨，略浸

- 50 g black fungi (wood ear):
  treated and cut into thick strips
- 50 g white fungi :
  refer to page 97 for its treatment and torn into pieces
- 1 tbsp Qi Zi :
  washed and soaked briefly

SeaSonIng 調味料

- 蒜茸 ....1 湯匙
- 芫茜 ....1 棵
  洗淨，切碎
- 鹽 ....1/4 茶匙
- 糖 ....1 湯匙
- 白醋 ....1 湯匙
- 生抽 ....1 湯匙
- 麻油 .... 少許

- 1 tbsp minced garlic
- 1 stalk coriander :
  washed and chopped
- 1/4 tsp salt
- 1 tbsp sugar
- 1 tbsp rice vinegar
- 1 tbsp light soy sauce
- dash of sesame oil

Method

做法

❶雙耳用滾水汆 2 分鐘，盛起；倒入杞子，一汆即取出，瀝乾水分，待涼。

❷把調味料拌勻，把 (1) 倒入，略醃，即可進食。

** 讓食材的調味更均勻，可將材料和調味料倒入膠袋內，搖勻略醃即可享用。

❶ Scald the double fungi in boiling water for 2 minutes and remove; pour in Qi Zi but remove it immediately after blanching. Drip dry and set it aside to cool.

❷ Mix well the seasoning, pour (1) in. Marinate briefly then serve.

** Put seasoning and ingredients in a plastic bag and shake them to mix well.

Cold Mixed Double Fungi

涼拌雙耳

Stir Fried Beancurd with Double Fungi

雙耳炒豆腐

InGreDienTs
材料

- 木耳 ....20 克
  處理好，切粗條
- 雪耳 ....20 克
  參閱第 97 頁處理法，再切成小朵
- 板豆腐 ....1 磚 切粒
- 腐乳 ....4 磚
  用少許腐乳汁加水攪爛，約半杯腐乳泥
- 糖 ....1 茶匙
- 蒜頭 ....4 粒 剁成茸
- 芫茜 ....2 棵 洗乾切碎
- 胡椒粉 .... 少許
- 麻油 .... 少許

- 20 g black fungi (wood ears):
  (treated and cut into thick strips)
- 20 g white fungi: refer to page 97 for its treatment and chopped
- 1 block beancurd : diced
- 4 pieces fermented beancurd:
  mashed with a little gravy and water to make 1/2 cup mashed fermented beancurd
- 1 tsp sugar
- 4 cloves garlic : minced
- 2 stalks coriander :
  washed and chopped
- pinch of pepper
- dash of sesame oil

MeThod
做法

❶用 1 湯匙油爆炒木耳及雪耳 2 分鐘，加鹽半茶匙，略兜炒即可盛起待用。
❷再起鑊，以 2 湯匙油先爆香蒜茸，倒入腐乳泥、豆腐粒，兜炒一會，再加入 (1)、胡椒粉和麻油，輕輕兜勻，下芫茜即成。

❶ Stir fry the two fungi in 1 tbsp of oil for 2 minutes. Add ½ tsp of salt and stir quickly. Scoop and set aside.
❷ Heat up the wok again. Fry minced garlic till fragrant in 2 tbsps of oil. Add the mashed fermented beancurd and diced beancurd to stir fry for a while. Put (1), pepper and sesame oil in to gently stir well. Sprinkle with chopped coriander to serve.

乾貨・雲耳
CLOUD EAR

挑選・SELECTING

宜挑選乾燥、手抓易碎、面呈烏黑光潤、底部略呈灰色像有絨面似的雲耳，如可以嘗少許，以味道清香的為佳。

It is good to buy those which are darkish black on surface and pick those which are dry and crumble easily when holding them in hands. Good cloud ear has darkish sheen on top and the bottom is grayish looking with crinkly surface. Choose those which are light and fragrant to the taste.

處理・TREATING

用清水浸軟及發大至兩三倍（約 1-2 小時），剪去蒂部的木屑，清洗乾淨即可烹調。

Soak them in water till softened and expanded 2-3 times their original size (about 1-2 hours). Cut off the woody bits at the stem; wash and clean them for use.

用途・USAGES

它亦是一個百搭材料，可滾湯、炒、燜或涼拌。用於食療，有潤膚養顏、益氣強身、治痔止血的功效。

It is a versatile ingredient with boiling soup, stir frying, braising and mixed cold dishes. Used as a healing food to moisten skin and strengthen our body. Also, it heals hemorrhoids and stops bleeding.

挑選 • SELECTING

宜挑選金黃色、較大粒、乾燥的黃耳，那些呈灰黑色的舊貨就不要購買了。

The golden color, bigger and dry ones are best. Do not buy the grayish and black old stock.

處理 • TREATING

用清水浸黃耳 12-24 小時，讓它充分發大，撕成細朵，剪去硬蒂，清洗乾淨後即可採用。

Soak yellow ear in water for 12-24 hours to allow them to expand fully. Tear into small pieces; cut off the stems, wash and ready for use.

用途 • USAGES

黃耳含豐富的膠質，滋潤而不膩滯，所以多用作湯、甜品和燜菜的材料。

They possess high jelly content which hydrates our body. Hence, they are often used in soup, desserts and braised dishes.

乾貨 • 黃耳 YELLOW EAR

Blessed Four Seasons

四季如意

InGreDienTs 材料

- 黃耳 ....19 克
  處理好，摘成小朵
- 雪耳 ....19 克
  參閱第 97 頁處理法，再摘成小朵
- 雲耳 ....19 克
  處理好，摘成小朵
- 榆耳 ....19 克
  浸過夜，洗淨，切塊
- 時菜 ....450 克
  洗淨後，摘去老葉

- 19 g yellow ears :
  treated and torn
- 19 g white fungi : refer to page 97
  for its treatment and torn
- 19 g cloud ears :
  treated and torn
- 19 g Yu ears : soaked overnight,
  washed and cut
- 450 g seasonal vegetable :
  washed and old leaves discarded

SeaSonIng 調味料

- 上湯 ....1 杯
- 糖 ....1 茶匙
- 鹽 .... 半茶匙
- 蠔油 ....2 湯匙

- 1 cup stock
- 1 tsp sugar
- 1/2 tsp salt
- 2 tbsps oyster sauce

MeThod 做法

❶用 1 杯水加 1 湯匙油、1 茶匙鹽焯時菜 2 分鐘，取出沖凍水，瀝乾水分鋪在碟邊。

❷用 2 片薑爆香四耳，加入上湯及調味料煮 15 分鐘，倒在焯熟的時菜上。

＊＊喜素食者，可將蠔油改為素蠔油，上湯改為素上湯，

以下是素上湯的材料、做法：

材料

大豆 .... 半斤

洗淨後用白鑊烘乾

羅漢果 ....1/4 個

紅蘿蔔 .... 1 條

去皮，切塊

西芹 ....2 條 切度

胡椒粒 ....1 湯匙

做法

用 6 杯水煮素上湯材料 1 個半小時，成 1 杯素上湯，待用。

❶ Blanch the vegetables in 1 cup of water plus 1 tbsp of oil and 1 tbsp of salt for 2 minutes; remove and splash them with cold water, drip dry and spread on the side of plate.

❷ Fry the four "ears" with 2 slices of ginger till fragrant. Add the seasoning to cook for 15 minutes. Pour on the blanched vegetables.

** For vegetarians, replace oyster sauce with the vegetarian version and change the stock to vegetarian stock.

Vegetarian Stock Ingredients

300 g soybean (washed and dry-fried in a wok)

1/4 piece Luo Han Guo

1 carrot (skinned and cut into pieces)

2 stalks celery (sectioned)

1 tbsp peppercorns

Method

Boil the ingredients with 6 cups of water for 1 1/2 hours to make 1 cup of vegetarian stock. Reserve for use.

乾貨・竹笙

ZHU SHENG

挑選 • SELECTING

一般在市場售賣的竹笙，價錢較平宜的，皆用塑膠袋包着售賣，不能用手觸它是否乾爽、味道是否清香，所以我們只能從顏色方面着手，宜挑選色澤淺黃，不要選太深色和雪白的，因為前者已放置長時間，而後者是經漂白的。
如有機會逐條挑選，應選氣味清香、乾爽和顏色淺黃的。

The Zhu Sheng (bamboo fungus) available at markets is normally sealed in plastic which make it impossible to feel their dryness and fragrance. Look at the color then, choose those which are light yellow but not too whitish or dark. The former has been bleached and the latter, has been on the shelf for too long. Given the opportunity to choose individually, select those fragrant, dry and yellowish ones.

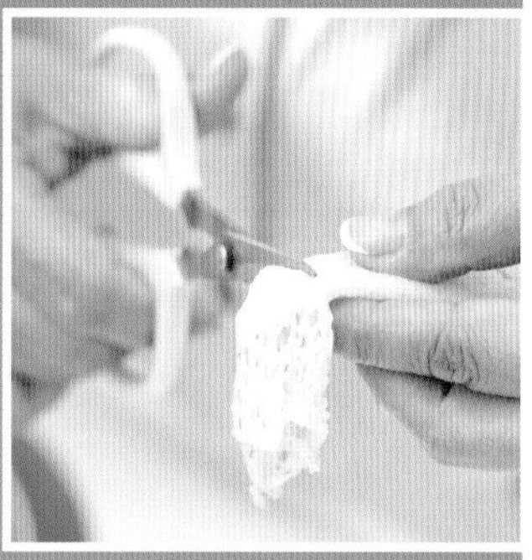

處理 • TREATING

竹笙用清水或洗米水浸軟，洗淨，剪去傘部和蒂，只要中段。
用洗米水浸竹笙，可去除竹笙的異味。

Soak Zhu Sheng (bamboo fungus) in water or water used for washing rice till they are soft. Cut off the stems and umbrella like tops, keep only the middle portions. Soaking them in used rice water can get rid of the smell.

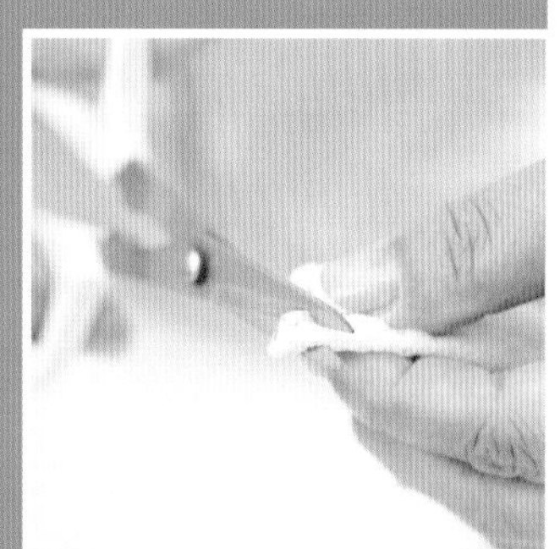

用途 • USAGES

多用來煲湯或配味鮮、味濃的材料共煮，因它味道不濃，吸味能力強。

Used mainly to boil soups or cook with fresh tasting and flavorful food, because its taste is light but it absorbs flavors well.

Steamed Prawns with Zhu Sheng

竹笙明蝦

InGreDienTs 材料

- 竹笙 ....12 條 處理好
- 中蝦 ....6 隻
  去殼，從背部切開一半，用鹽抓洗兩次
- 蘆筍 ....3 條 只要嫩莖
- 洋火腿 ....1 片
  切開 12 小條
- 冬菇 ....4 朵
  參閱第 80 頁處理法，再每朵切開 3 條
- 芫茜 .... 少許

- 12 Zhu Sheng : treated
- 6 prawns : shells removed. Slit into half. Washed and rubbed with salt twice
- 3 asparagus : keeping only the tender sections
- 1 slice ham : cut into 12 portions
- 4 dried shiitake mushrooms : refer to page 80 for it treatment and then cut into 3 strips
- coriander

Sauce 獻汁

- 上湯 ....1 杯
- 粟粉 ....2 茶匙
- 鹽 .... 半茶匙
- 麻油 ....1 茶匙
- 胡椒粉 .... 少許

- 1 cup stock
- 2 tsps cornflour
- 1/2 tsp salt
- 1 tsp sesame oil
- pinch of pepper

MeThod 做法

❶將蘆筍、火腿、蝦及冬菇小心塞進竹笙內，猛火蒸 7 分鐘。
❷把竹笙排好，埋獻淋在竹笙上，飾上芫茜即成。

❶ Stuff the asparagus, ham, prawn and mushrooms carefully into the Zhu Sheng. Steam over high heat for 7 minutes.
❷ Arrange Zhu Sheng neatly, pour sauce over them. Garnish with coriander to serve.

Steamed Prawns with Zhu Sheng

竹笙明蝦

乾貨・陳皮

DRIED TANGERINE PEEL

挑選・SELECTING

話明是陳皮，即是年份越久的果皮就越珍貴，味道越甘醇。購買時宜挑選色澤深、香味濃、乾爽的陳皮。有些不法的奸商將乾柑皮染色以冒充陳皮，所以宜往有信譽的店舖購買。

The more aged the tangerine peel, the more fragrant it gets. Pick those which are darker in color, more fragrant and dry. As some unscrupulous merchants dye the color of orange peel to pass off as dried tangerine peel, do buy them at reputable shops.

處理・TREATING

陳皮用温水浸軟後刮去瓤，即可用作烹調；但年份久的陳皮是不用去果瓤的，因果瓤已變得非常薄。

For cooking, soak them in warm water to soften before scrapping off the pith. However, aged dried tangerine peel has very thin pith hence it is unnecessary to scrape it off.

用途・USAGES

陳皮雖然是配角，但對食物有畫龍點睛之效，它宜蒸、燉、焖、煲湯、煮甜品等。

It is great for enhancing the flavor of food. It can be used in steaming, braising, boiling soup and making desserts.

Veal Ribs with Dried Tangerine Peel

陳皮牛仔骨

InGreDienTs 材料

- 陳皮 .... 半個 處理好，切絲
- 牛仔骨 ....1 磅 (450 克)
- 乾紅椒 ....2 隻
- 花椒 ....30 粒
- 乾葱 ....2 粒 切片
- 薑 ....2 片
- 紹酒 ....1 湯匙

- 1/2 dried tangerine peel : treated and shredded
- 450 g veal ribs
- 2 dried red chilies
- 30 Sichuan peppercorns
- 2 shallots : sliced
- 2 slices ginger
- 1 tbsp Shaoxing wine

MariNade 醃料

- 生抽 ....1 湯匙
- 糖 ....1 茶匙
- 粟粉 ....1 茶匙
- 胡椒粉 .... 少許

- 1 tbsp light soy sauce
- 1 tsp sugar
- 1 tsp cornflour
- pinch of pepper

MeThod 做法

❶牛仔骨先用醃料醃半小時，再用油煎香兩面。

❷用 2 湯匙油爆香花椒、陳皮絲、乾紅椒、乾葱、薑片，加牛仔骨回鑊，兜炒，灒酒，兜勻上碟。

❶ Marinate the veal ribs for 30 minutes, then fry the two sides till fragrant in some oil.

❷ Fry the Sichuan peppercorns, dried tangerine peel shreds, dried red chilies, spring onion, ginger and shallots in 2 tbsps of oil till fragrant. Add the ribs to the wok to stir quickly. Pour wine on the side of wok, stir well and serve.

乾貨・杞子 QI ZI

挑選 • SELECTING

宜挑選乾身、色澤鮮明、沒有蟲蛀，細嚼時味道甘甜的杞子。

Choose those which are dry, bright in color and without worms. They should be sweet to the taste.

處理 • TREATING

沖洗後略浸一會至軟，就可採用了。

After rinsing, soak them briefly till softened.

用途 • USAGES

多用來煲湯、蒸餸、焗茶。

Used mainly to boil soup, in steamed dishes and making tea.

Stir Fried Lily Bulbs with Asparagus

百合扒蘆筍

InGreDienTs 材料

- 新鮮百合 ....1 個
  剝開每一瓣，沖洗乾淨
- 蘆筍 .... 半斤
  折去老的部分，一開四
- 杞子 ....2 湯匙
  處理好

- 2 fresh lily bulbs :
  peel each petal to wash
- 300 g asparagus :
  discard the old sections, halved
- 1 tbsp Qi Zi : treated

SeaSonIng 調味料

- 雞湯 .... 半杯
- 粟粉 ....1 茶匙
- 鹽 .... 半茶匙
- 糖 ....1/4 茶匙
- 胡椒粉 .... 少許
- 麻油 .... 少許

- 1/2 cup chicken soup
- 1 tsp cornflour
- 1/2 tsp salt
- 1/4 tsp sugar
- pinch of pepper
- dash of sesame oil

Stir Fried Lily Bulbs with Asparagus

百合扒蘆筍

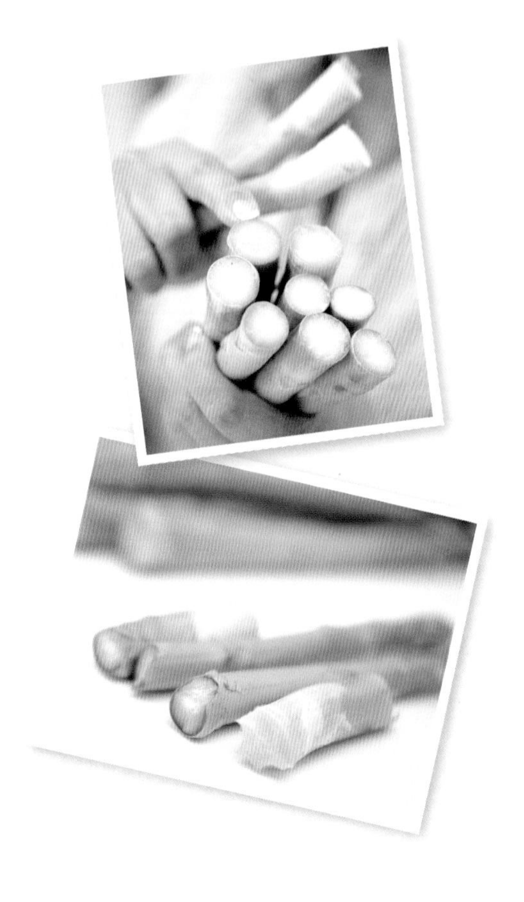

Method

做法

❶煲滾 2 杯水，加 1 湯匙油、1 茶匙鹽，放入蘆筍，約 2 分鐘，撈起，排放在碟上。用同一鍋水，汆百合和杞子約 10 秒，撈起排在蘆筍上。
❷用細火把調味料混合煮開，即可淋（1）上。
＊＊ 市場上有抽真空包裝的新鮮百合出售，也有已焙乾、曬乾的乾貨，鮮貨以略具光澤為佳。
＊＊挑選蘆筍的竅門：在超市購買的蘆筍多是一紮紮的，想知道它是否嫩，就要看它的切口了。如切口呈翠綠色代表蘆筍非常鮮嫩，相反如呈白色則代表蘆筍已老、吃時會有渣，處理時須削去老梗。

❶ Bring 2 cups of water to the boil; add 1 tbsp of oil, 1 tsp of salt to blanch the asparagus for 2 minutes. Scoop and arrange them on a plate. Then, blanch the lily bulbs and Qi Zi for 10 seconds in the same pot. Scoop and place on top of the asparagus.
❷ Cook the ingredients for the seasoning over low heat. Pour over (1) to serve.
** Lily bulbs are available in fresh vacuum-packed packages or in the sun-dried and roasted forms. Select those with bright sheen.
** Check for the freshness of stalks of asparagus available at supermarkets by looking at their edge. If the color is bright green they are fresh. On the contrary those showing whitish color are old and coarse in taste. Do peel off the old and hard stems.

Steamed Qi Zi with Eggs

杞子蒸蛋

InGreDienTs 材料

- 蛋........4 個
  打成蛋液
- 水 蛋液的 1.5 倍
- 杞子........1 湯匙
  處理好，用廚房紙印乾水分
- 鹽........1 茶匙

- 4 eggs : beaten
- water : about 1.5 times the volume of egg liquid
- 1 tbsp Qi Zi : treated and absorted water with kitchen paper
- 1 tsp salt

MeThod 做法

❶蛋液和水調勻，下鹽，拌勻，篩去泡沫。
❷煲滾一鍋水，放入蒸架和碟，蒸至碟熱透，倒入蛋液，放上杞子，蓋上鍋蓋，隔熱水猛火蒸兩分鐘。
❸關火，焗 20 分鐘後開蓋，可下生抽、灒熟油伴食。

＊＊杞子沖淨即可，不需要浸泡，否則杞子會沉底。

＊＊擁有我第一本食譜的朋友，可能已試過書內的「嫩滑水蛋」，今次我略加變化，加入杞子同蒸，讓菜式添上新意。

❶ Mix the eggs with water well; add salt, stir and scrape off the surfaced bubbles.
❷ Bring a pot of water to the boil; place the steaming stand and a plate on it to heat up the plate. Pour the egg liquid in, add the Qi Zi. Cover and steam over high heat for 2 minutes.
❸ Switch off flame and leave for 20 minutes before lifting the cover. Add light soy sauce and cooked oil to serve.

** Wash Qi Zi without soaking them, or they may sink to the bottom of dish.

** For those of you who have my first cookbook, you probably would have tried the dish, "Steamed Egg Custard". I have added Qi Zi for a new taste.

Huai Shan and Qi Zi Porridge

淮山杞子粥

InGreDienTs
材料

- 淮山粉 ....38 克
  可請藥材舖代磨
- 杞子 ....19 克
  處理好
- 圓米 .... 半杯
- 水 ....8 杯

- 38 g ground Huai Shan: ground it at medicinal shops
- 19 g Qi Zi : treated
- ½ cup short-grained rice
- 8 cups water

MeThod
做法

❶用 1 杯凍水調勻淮山粉。
❷大火煲滾 8 杯水，加入米及淮山粉水，滾後改用中小火煲約 1 小時，加入杞子，再煲 30 分鐘即成。
＊＊日本米和蓬萊米都屬圓米。

❶ Mix the Huai Shan powder well in a glass of cold water.
❷ Bring 8 cups of water to the boil, add the rice and Huai Shan water. Then, turn it down to medium low heat to boil for 1 hour, add Qi Zi. Boil for another 30 minutes to serve.
** Both Japanese and Penglai rice are short-grained rice.

乾貨 • 花生

PEANUTS

挑選 • SELECTING

要挑選顆粒飽滿、沒有蟲蛀、乾爽、細嚼時滿嘴花生香、沒有霉味的花生。

The dry, round and plump ones without worms are the best. When eating them, their flavor should be full of peanut fragrance but not moldy in taste.

處理 • TREATING

如花生是用來油炸或烘焙，用水沖去花生表面的塵埃，再風乾即可使用。如用來煲湯，宜浸 1-2 小時。

Rinse off the surface dust of peanuts before drying them, if they are used for deep-frying or roasting. For making soup, soak them in water for 1-2 hours.

用途 • USAGES

花生的用途廣泛，可用來做甜品、小食、燜肉、湯羹等等。

Peanuts have many uses such as making desserts, snacks, casserole or soup.

Peanut Cookies

花生餅

InGreDienTs
材料

- 花生........150 克
  處理好
- 花生醬........4 湯匙
- 糖霜........2 湯匙
- 牛油........2 湯匙
- 糕粉........4 湯匙

- 150 g peanuts : treated
- 4 tbsps peanut butter
- 2 tbsps icing sugar
- 2 tbsps butter
- 4 tbsps steamed glutinous rice flour

* 花生餅模

* Peanut cookie mould

花生餅

Method

做法

❶預熱焗爐 175℃。

❷花生用白鑊慢火炒香（約 10 分鐘），盛起，略凍後放入膠袋內搓去衣；將花生肉磨或舂成幼粒。

❸把（2）連同花生醬、糖霜和牛油搓勻，放入糕粉搓勻，再按你的糕模大小分成等份。

❶ Preheat the oven to 175℃ .

❷ Fry the peanuts in a dry wok till fragrant (about 10 minutes). Scoop, let it cool somewhat then place in a plastic bag to remove husks; pound it finely.

❸ Rub (2) with the peanut butter, icing sugar and butter well. Mix with the steamed glutinous rice flour. Press the filling into the peanut cookie moulds to shape.

Peanut Cookies

花生餅

Method

做法

❹餅模內灑上糕粉，拍去多餘糕粉，把（3）按實在餅模內，略敲餅模的四邊（以便易於反扣），反扣，花生餅排在已灑上糕粉的焗盤上。

❺將焗盤放進焗爐內焗 10 分鐘即成。

＊＊ 糕粉在烘焙材料店有售。

＊＊ 糕粉你亦可自製：先用猛火蒸 10 分鐘，再用白鑊細火炒 8 分鐘即成，用剩的糕粉可貯在密封瓶內。

❹ Sprinkle the steamed glutinous rice flour on the moulds, remove the excessive flour. Press (3) into the moulds, press the edges of the moulds slightly (for easier inverting). Invert them to place on the floured baking tray.

❺ Bake in the preheated oven for 10 minutes. Serve.

** You may buy steamed glutinous rice flour from any baking supply stores.

** Or, the homemade version: Steam glutinous rice flour over high heat for 10 minutes, and then fry in a dry wok over low heat for 8 minutes. Store the unused flour in a tightly sealed jar.

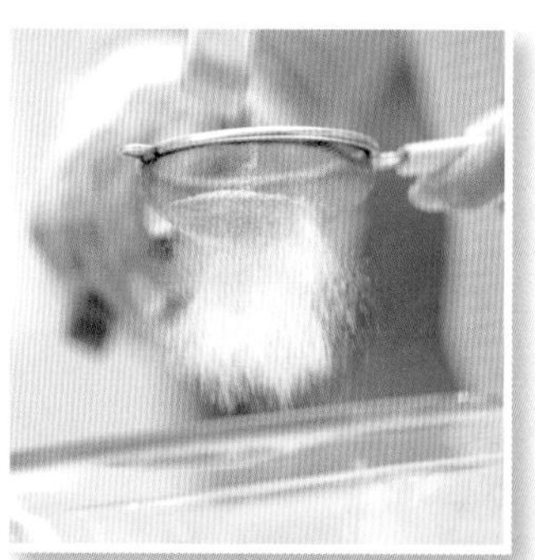

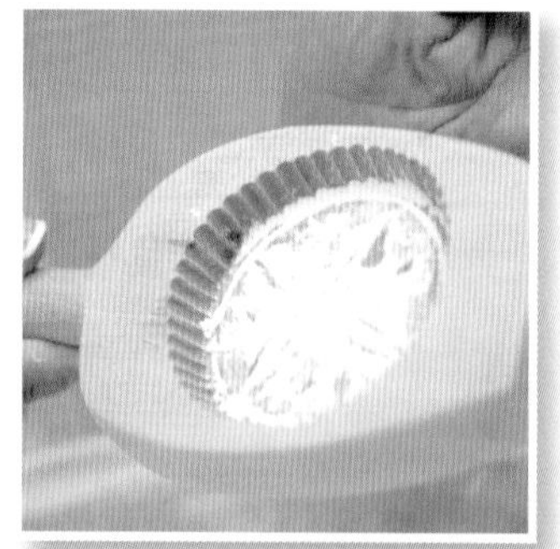

乾貨・芡實 FOX NUTS

挑選・SELECTING

芡實以顆粒飽滿、完整、沒有「糠」味和蟲蛀的為佳。

Pick the round and plump ones intact and with no bad odor nor worm-infested.

處理・TREATING

挑去雜質，略洗，用清水浸1-2小時(視乎芡實的用途)，即可使用。

Pick out and discard the impurities and rinse briefly. Soak in water for 1-2 hours depending on its usage.

用途・USAGES

煲湯、煲粥、煮糖水。

Use it in boiling soup and porridge, plus for making sweet soup.

註・REMARKS

因為拍攝「吾湯吾水」的關係，讓我有機會看到芡實的真面目。一粒粒躺在葉上的果實，皮堅且硬，肉地潔白呈粉性。我喜歡它那QQ的口感，而且它非常有益，有補脾止泄的食療功效。

Over the filming of "The Secrets of Soup", I had the opportunity to look at fox nuts up close. Lying individually on the leaves, the fruits have firm and hard skin with white and powdery flesh. I love its chewy texture when biting it. Moreover, fox nuts improve the sheen of our skin and stop diarrhea.

Fox Nuts and Mature Duck Soup

芡實老鴨湯

InGreDienTs 材料

- 芡實 ....30 克
  處理好
- 淮山 ....30 克
  沖洗後浸片刻
- 老鴨 ....1 隻約 2 千克
  洗淨去內臟，汆水
- 瘦肉 ....600 克 汆水
- 老薑 ....2 厚片 拍鬆
- 葱 ....2 條 打結
- 水 ....18 杯

- 30 g fox nuts : treated
- 30 g Huai Shan : rinsed and soaked briefly
- 1 mature duck about 2 kg : washed and innards discarded, scalded
- 600 g lean pork : scalded
- 2 thick slices ginger : pat to loosen
- 2 stalks spring onion : tied in knots
- 18 cups water

MeThod 做法

❶將所有材料放入鍋內，先用大火煲至滾，10 分鐘後收至中小火煲 3 小時即可。
❷加鹽 1 茶匙調味。

❶ Place all the ingredients in a pot; bring to the boil over high heat. After 10 minutes, lower to medium heat to boil for 3 hours.
❷ Season with 1 tsp of salt.

Lotus Seeds and Fox Nuts Sweet Soup

蓮子芡實糖水

InGreDienTs
材料

- 蓮子 ....1/4 杯
  用 2 杯水浸 2 小時，叮 20 分鐘
- 芡實 ....1/4 杯
  處理好，再叮 10 分鐘
- 薏米 ....1/4 杯
  用 2 杯水略浸， 叮 5 分鐘
- 龍眼肉 ....10 克 略沖洗
- 滾水 ....6 杯
- 蜂蜜 ....3 湯匙

- 1/4 cup lotus seeds :
  soaked in 2 cups of water for 2 hours and microwaved for 20 minutes
- 1/4 cup fox nuts :
  treated and microwaved for 10 minutes
- 1/4 cup Job's tears :
  soaked briefly in 2 cups of water and microwaved for 5 minutes
- 10 g shelled dried longans :
  rinsed briefly
- 6 cups boiling water
- 3 tbsps honey

MeThod
做法

❶ 把叮過的材料和連水放入煲內，另加 6 杯滾水和龍眼再煲約 30 分鐘，加入蜂蜜，
即可享用。

＊＊利用微波爐「叮」熟乾貨及豆類，可以節省能源及時間，而且更能控制材料質感。

＊＊此糖水煲好時會有浮泡，撇走即可。

❶ Pour all the microwaved ingredients into a pot; add 6 cups of boiling water and longans to cook for about 30 minutes. Add the honey to serve.

** Using microwave to cook dried goods and pulses could save a lot of time and energy. Plus, it is better in preserving the textures of ingredients.

** Bubbles will emerge on top of this sweet soup when it is boiled. Do scoop it away before serving.

挑選 • SELECTING

從表面看，宜挑選顆粒飽滿、表皮皺紋少、顏色暗紅、沒破損、不過度乾癟的；但你也需要用手按一按，如棗肉太軟也不要買，因為可能曬的日子不足，內裏已發霉或蟲蛀。

Pick those round and plump, less crinkled, dark red, unbroken and less dry ones. However, do touch it to feel the texture, don't buy the ones which feel too soft as they may not be sundried enough, hence they are rotten or worm-infested.

處理 • TREATING

沖洗、去塵埃，用刀背拍扁，去核（據說紅棗核帶燥，宜去掉才煲湯）。

Wash and clean off the dust. Use the back of a knife to flatten them and to remove the stones before using them (it is said that red dates with stones are heaty, it is best to remove them for boiling soup).

用途 • USAGES

多用來煲湯、蒸和製作中式糕點。

Used mainly to boil soup or steaming and making Chinese style cakes.

Red Date Cake

紅棗糕

InGreDienTs

材料

- 河南棗 ....300 克
- 蛋 ....8 個
  蛋白、蛋黃分開盛起
- 糖 ....1/3 杯
- 自發粉 ....1 杯
  篩兩次
- 菜油 .... 半杯
- 紅棗水 .... 半杯

- 300 g Henan red dates
- 8 eggs :
  yolks and whites separate
- 1/3 cup sugar
- 1 cup self-raising flour
  sifted twice
- ½ cup vegetable oil
- ½ cup red dates water

Red Date Cake

紅棗糕

Method

做法

❶用3杯半水將紅棗煲滾，再轉慢火煲1小時，倒起紅棗水待用，紅棗則去皮、去核，取肉待用。

❷蛋白打至企身，約6分鐘。

❸蛋黃、紅棗水、糖及菜油用電動打蛋器打2分鐘，加入紅棗泥打勻，改用慢速逐少加入自發粉。

❹將（2）倒入（3）中，輕手拌勻。

❺蒸盤先抹菜油，倒入（4）。

❻猛火蒸45分鐘即成。

＊＊ 河南棗體積較大，比較容易去皮。

＊＊ 紅棗去皮花費時間，所以要預先準備紅棗泥。

＊＊ 紅棗糕有啖啖甜美的棗香，口感輕軟。

❶ Bring the red dates in 3 ½ cups of water to the boil. Then turn down the heat to low to cook for 1 hour. Pour and keep the red dates water for use later. Remove the skin and piths of the red dates, keep the flesh.

❷ Beat the egg whites till soft peak form for about 6 minutes.

❸ Beat the egg yolks and add the red dates water, sugar and oil with an electric mixer for 2 minutes. Add mashed red dates to mix well, change to slow speed to gradually add the self-raising flour.

❹ Pour (2) into (3) to fold gently with spatula.

❺ Grease a steaming tray with the vegetable oil, pour (4) in.

❻ Steam on high heat for 45 minutes to serve.

** Henan red dates are larger; hence it is easier to remove the skins.

** Removing the skins take some time. Therefore prepare the mashed red dates early.

** Red date cake is fragrant and has nice texture.

Red Dates and Longan Drink

紅棗桂圓茶

InGreDienTs
材料

- 紅棗 ....20 粒
  處理好
- 桂圓 ....20 粒
  略沖洗
- 水 ....6 杯

- 20 red dates :
  treated
- 20 shelled dried longans :
  rinsed briefly
- 6 cups water

MeThod
做法

❶紅棗、桂圓和水一起放入煲內，先用大火煲滾，然後轉調慢火煲 30 分鐘，即可享用。

＊＊ 這茶水在臨睡前飲用，有安神的功效。

❶ Place red dates, longans and water in a pot, bring to the boil then turn the flame to low to boil for 30 minutes. Serve.

** Drink this tea before bed, it soothes the nerves.

Calming Red Dates Soup

紅棗安神湯

InGreDienTs 材料

- 百合 ....3 錢
  沖洗後略浸
- 烏雞 ....1 隻
  洗淨，汆水
- 何首烏 ....3 錢
  沖洗後略浸
- 茯神 ....3 錢
  沖洗後略浸
- 酸棗仁 ....3 錢
  沖洗後用刀背拍碎
- 紅棗 ....12 粒
  處理好
- 水 ....10 杯

- 12 g dried lily bulbs : washed and soaked briefly
- 1 black-skinned chicken : washed and scalded
- 12 g He Shou Wu : washed and soaked briefly
- 12 g Fu Shen : washed and soaked briefly
- 12 g Suan Zao Ren : washed and smashed with the back of knife
- 12 red date : treated
- 10 cups water

MeThod 做法

❶除百合及烏雞，所有材料放入煲湯袋內。
❷煲滾 10 杯水，放入所有材料，用大火滾 5 分鐘後改用小火再煲 2 小時。
❸飲前下 1 茶匙鹽調味即可。
＊＊ 酸棗仁有改善失眠的功效，配以百合與茯神，是極佳的安神湯。

❶ Put all the ingredients except the black-skinned chicken and lily bulbs into a muslin bag.
❷ Bring 10 cups of water to the boil; pour all ingredients into a pot. Bring it to the boil for 5 minutes, turn to low heat to boil for another 2 hours.
❸ Before serving, season it with 1 tsp of salt.
** Suan Zao Ren relieve insomnia. Adding it to lily bulbs and Fushen, it becomes an excellent nerve-calming soup.

InGreDienTs
材料

- 沙參 ....75 克 洗淨
- 玉竹 ....75 克 洗淨
- 麥冬 ....38 克
- 瘦肉 ....380 克
  汆水，切大件
- 雞 ....1 隻約 900 克
  劏洗乾淨，汆水，斬件
- 紅棗 ....6 粒
  處理好
- 水 ....16 杯

- 75 g Sha Shen : rinsed
- 75 g Yu Zhu : rinsed
- 38 g Mai Dong
- 380 g lean pork :
  scalded and cut into chunks
- 1 chicken about 900 g
  washed and cut into pieces
- 6 red dates : treated
- 16 cups water

MeThod
做法

❶煲滾水，放入所有材料，用大火煲滾後，改用小火煲 3 小時，下少許鹽調味即可飲用。

** 沙參、玉竹有相輔相成的作用，可潤燥、除煩、生津止渴。

❶ Bring water to the boil, add all the ingredients. Boil over high heat then turn to low heat to boil for 3 hours. Season it with a little salt.

** Sha Shen and Yu Zhu are complementary food for cooling nerves and quenching thirst.

Sa Shen, Yu Zhu and Chicken Soup

沙參玉竹雞湯

乾貨 • 羅漢果

LUO HAN GUO

挑選 • SELECTING

應挑選外殼黃褐色、完整沒破損、搖不響的羅漢果；如搖時有聲響，表示羅漢果放置過久內裏已有蟲蛀。

The intact unbroken ones showing yellow-brownish color on their shells are the choice. When shaking the fruit, those with "sound" have been left on shelves for too long and are worm-infested.

處理 • TREATING

洗淨果實表面塵污即可。

Rinse off the surface dirt.

用途 • USAGES

多用來煲老火湯及茶水。羅漢果的甜度高，注意使用的份量。

Used mainly for long-boiled soup and tea. Careful of the quantity to use as its sugar content is high.

Luo Han Guo and Loquat Leaves with Pig's Lungs Soup

羅漢果枇杷葉豬肺湯

InGreDienTs 材料

- 銀杏 ....100 克
  去殼、去衣
- 新鮮枇杷葉 ....5 片
- 羅漢果 .... 半個
  處理好
- 豬肺 ....1 個
  從豬氣管灌水至豬肺，
  一邊用手擠出豬肺內的血污，
  直至水變白，再擠去水分。
- 瘦肉 ....250 克
  汆水
- 蜜棗 ....4 粒
- 南北杏 ....2 湯匙
- 水 ....18 杯

- 100 g gingko nuts : shells removed and skinned
- 5 pieces fresh loquat leaves
- ½ Luo Han Guo : treated
- 1 pig's lungs : Fill the lungs from the air ducts with water. Use hands to squeeze the blood impurities from the lungs till the water turns whitish; squeeze out the water again.
- 250 g lean pork : scalded
- 4 candied dates
- 2 tbsps bitter and sweet almonds
- 18 cups water

Luo Han Guo and Loquat Leaves with Pig's Lungs Soup

羅漢果枇杷葉豬肺湯

Method

做法

❶將灌水後的豬肺切成大塊，放在白鑊中烘乾。

❷所有材料放入滾水內，用中慢火煲 3 小時即成。

＊＊ 若買不到新鮮枇杷葉，可到中藥房買乾貨代替。

＊＊ 豬肺宜切成大塊才處理，因經白鑊烘過的豬肺會縮小一半。

＊＊ 豬肺灌水的步驟，可請相熟的肉檔代勞。

❶ Slice the rinsed pig's lungs into pieces, dry them in a dry and hot wok.

❷ Place all the ingredients into boiling water; boil over medium low heat for 3 hours to serve.

** Replace the fresh loquat leaves with the dried version obtainable from Chinese medicinal shops, if they are not available.

** Cut pig's lungs into chunks before treating it, as the cooked lungs will shrink in half.

** The job of rinsing pig's lungs can be done by pork vendors.

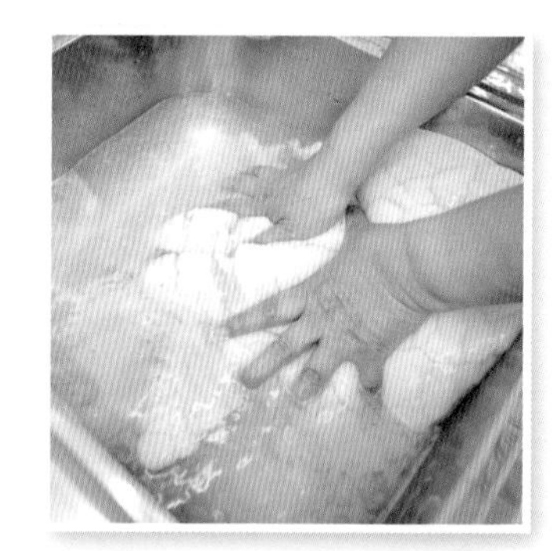

Luo Han Guo and Pear Drink

羅漢果雪梨水

InGreDienTs
材料

- 羅漢果........半個
  處理好
- 雪梨........2 個
- 水........6 杯

- 1/2 Luo Han Guo : treated
- 2 Chinese pears
- 6 cups water

MeThod
做法

❶羅漢果、雪梨和水一起放入煲內，先用大火煲滾，然後轉調慢火煲 30 分鐘，即可享用。

** 這茶水有生津、潤肺的功效。

❶ Place the Luo Han Guo, Chinese pears and water into a pot. Bring to the boil over high heat. Then turn it down to low flame to boil for 30 minutes. Serve.

** This drink quenches thirsts and it is soothing to lungs.

乾貨・合桃 WALNUTS

挑選 • SELECTING

市場上售賣的合桃，大致可分為兩種，一是連殼的，二是淨肉。購買有殼的合桃時，宜挑選外殼完整、鮮褐色帶光澤的；因外殼破損的合桃，果仁的味道可能已有變，而果殼較為白淨的，可能經過漂白。
而挑選淨合桃肉要注意的，是看看是否有蟲蛀和是否有「油益」味。

There are two types of walnuts available at marketplace: with or without shells. When buying the shelled ones, choose those with shells intact and bright brownish in color with sheen to it. If shells are broken, it may mean that the taste of the nuts have gone rancid. However, if shells are too white, it could mean they have been bleached.
Look out for worms or rancid smell when you choose the unshelled walnuts.

處理 • TREATING

略沖去塵埃，再風乾即可使用。合桃衣的營養價值高，不要撕去。

Rinse off the surface dust of walnuts before drying them. The membranes of walnuts possess good nutrients which should not be discarded.

用途 • USAGES

合桃的用途繁多，可整顆炸或焗成小食，可研糊炮製成甜品，亦可用來煲湯、燉湯。

Use it in a great variety of ways – fry the whole nuts or roast them as snacks; make them into a paste for desserts or use it for boiling soup.

Walnut Cakes

合桃蛋糕

InGreDienTs
材料

- 合桃 ....2 杯半
  處理好
- 糖 ....3/4 杯
- 蛋 ....6 個
- 麵粉 .... 半杯
- 牛油 ....5 湯匙
  室溫
- 檸檬皮 ....1 湯匙
  磨茸
- 糖霜 .... 適量

- 2 1/2 cups walnuts :
  treated
- 3/4 cup sugar
- 6 eggs
- 1/2 cup flour
- 5 tbsps butter :
  at room temperature
- 1 tbsp lemon zest :
  grated
- some icing sugar

MeThod
做法

❶預熱焗爐 175℃ (350F) 。
❷合桃與糖同放攪拌機攪碎，放在大碗備用。
❸將蛋逐個加入合桃碎內混合，再順序加入麵粉、牛油及檸檬皮茸，拌勻，倒入 12 個已抹油的模內。
❹放入已預熱的焗爐內焗 30 分鐘，取出，待略涼後灑上糖霜。
＊＊ 可伴上蜂蜜享用。

❶ Preheat oven to 175℃ (350F).
❷ Combine the walnuts and sugar in a mixer to crush, place in a big bowl for later use.
❸ Mix the egg one by one into the mashed walnuts, and in this order add flour, butter and lemon zest. Pour into 12 greased moulds.
❹ Place in the preheated oven to bake for 30 minutes. Remove. Let cool then sprinkle the icing sugar on top to serve.
** Enjoy it with honey!

Walnut Cakes

合桃蛋糕

手機一嘟即時去片！
http://e.formspub.com/videos/?id=hamtaocake

Walnut and Sweet Ginger Drink

胡桃仁薑糖飲

InGreDienTs

材料

- 合桃 ....1 杯
  略沖洗
- 薑 ....5 片
  剁成薑米
- 紅糖 ....1/2 杯
- 水 ....4 杯

- 1 cup walnuts :
  washed briefly
- 5 slices ginger :
  diced
- 1/2 cup brown sugar
- 4 cups water

MeThod

做法

❶煲滚水，放入所有材料，用大火煲約15 分鐘即可飲用。

＊＊ 這茶飲對感冒、頭痛和發燒有一定療效。

❶ Bring the water to the boil, place all the ingredients in and boil over high heat for 15 minutes before serving.

** It helps in eliminating common colds, headache and fever.

海味。乾貨。家常菜

歡迎加入 Forms Kitchen「滋味會」

登記成為「滋味會」會員

• 可收到最新的飲食資訊 •

• 書展"驚喜電郵"優惠 * •

• 可優先參與 Forms Kitchen 舉辦之烹飪分享會 •

• 您喜歡哪類飲食叢書？(可選多於 1 項)

☐中菜 ☐西菜 ☐點心 ☐烘焙 ☐湯飲 ☐甜品 ☐其他____________

• 您對哪類飲食題材感興趣，而坊間未有出版品提供，請說明：

____

• 此書吸引您的原因是：(可選多於 1 項)

☐興趣 ☐內容豐富 ☐封面吸引 ☐工作或生活需要

☐作者 ☐價錢相宜 ☐其他____________

• 您從何途徑擁有此書？

☐書展 ☐報攤 / 便利店 ☐書店 (請列明：____________)

☐朋友贈予 ☐購物贈品 ☐其他____________

• 您覺得此書的價格：

☐偏高 ☐適中 ☐因為喜歡，價錢不拘

• 除食譜外，您喜歡閱讀哪類書籍？(可選多於 1 項)

☐玄學 ☐旅遊 ☐心靈勵志 ☐健康美容 ☐語言學習 ☐小說

☐兒童圖書 ☐家庭教育 ☐商業創富 ☐文學 ☐宗教

☐其他____________

• 您是否有興趣參加作者的烹飪分享活動？

☐有興趣 ☐沒有興趣

• 哪位作者的烹飪分享活動您會有興趣參加？

____

姓名：____________ ☐男 / ☐女 ☐單身 / ☐已婚

聯絡電話：____________ 電郵：____________

地址：____

年齡：☐ 20 歲或以下 ☐ 21-30 歲 ☐ 31-45 歲 ☐ 46 歲或以上

職業：☐文職 ☐主婦 ☐退休 ☐學生 ☐其他____________

填妥資料後可：

寄回：香港鰂魚涌英皇道 1065 號東達中心 1305 室「Forms Kitchen」收

或傳真至：(852) 2565 5539

或電郵至：info@wanlibk.com

*請 ✔ 選以下適用的項目

☐我已閱讀並同意萬里機構出版有限公司訂立的《私隱政策》聲明# ☐我希望定期收到新書及活動資訊

有關使用個人資料安排

您好！為配合《2012 年個人資料（私隱）（修訂）條例》（《修訂條例》）的實施，包括《2012 年個人資料（私隱）（修訂）條例》中的第 2(b) 項，萬里機構出版有限公司（下稱"本社"）希望閣下能充分了解本社使用個人資料的安排。

為與各曾跟萬里機構出版有限公司接觸的人士及已招收的會員保持聯繫，並讓閣下了解本社的最新消息，包括新書推介、會員活動邀請、推廣及折扣優惠訊息、問卷調查、其他文化資訊及收集意見等，本社會不時向各位發放相關信息。本社會使用您的個人資料（包括姓名、電話、傳真、電郵及郵寄地址），來與您繼續保持聯繫。

除作上述用途外，本社將不會將閣下的個人資料以任何形式出售、租借及轉讓予任何人士或組織。

海味。乾貨。家常菜

作者 Author
黃淑儀 Gigi Wong

策劃/編輯 Project Editor
Catherine Tam

翻譯 Translator
Patricia Mok

攝影 Photographer
Imagine Union

美術統籌及設計 Art Direction & Design
Amelia Loh

出版者 Publisher
Forms Kitchen
香港鰂魚涌英皇道1065號東達中心1305室
Room 1305, Eastern Centre, 1065 King's Road, Quarry Bay, Hong Kong.
電話 Tel: 2564 7511
傳真 Fax: 2565 5539
電郵 Email: info@wanlibk.com
網址 Web Site: http://www.wanlibk.com
http://www.facebook.com/wanlibk

發行者 Distributor
香港聯合書刊物流有限公司
SUP Publishing Logistics (HK) Ltd.
香港新界大埔汀麗路36號中華商務印刷大廈3字樓
3/F., C&C Building, 36 Ting Lai Road, Tai Po, N.T., Hong Kong
電話 Tel: 2150 2100
傳真 Fax: 2407 3062
電郵 Email: info@suplogistics.com.hk

承印者 Printer
中華商務彩色印刷有限公司
C & C Offset Printing Co. Ltd.

出版日期 Publishing Date
二零一一年六月第一次印刷
First print in June 2011
二零一九年三月第五次印刷
Fifth print in March 2019

Published in Hong Kong
ISBN 978-988-8295-10-4

鳴謝：
阿一鮑魚始創人楊貫一先生
Acknowledgement :
Founder of Ah Yat Abalone, Yeung Koon Yat